POOLING DESIGNS AND NONADAPTIVE GROUP TESTING

Important Tools for DNA Sequencing

SERIES ON APPLIED MATHEMATICS

Vol. 1 International Conference on Scientific Computation
eds. T. Chan and Z.-C. Shi

Vol. 2 Network Optimization Problems — Algorithms, Applications and Complexity
eds. D.-Z. Du and P. M. Pandalos

Vol. 3 Combinatorial Group Testing and Its Applications
by D.-Z. Du and F. K. Hwang

Vol. 4 Computation of Differential Equations and Dynamical Systems
eds. K. Feng and Z.-C. Shi

Vol. 5 Numerical Mathematics
eds. Z.-C. Shi and T. Ushijima

Vol. 6 Machine Proofs in Geometry
by S.-C. Chou, X.-S. Gao and J.-Z. Zhang

Vol. 7 The Splitting Extrapolation Method
by C. B. Liem, T. Lü and T. M. Shih

Vol. 8 Quaternary Codes
by Z.-X. Wan

Vol. 9 Finite Element Methods for Integrodifferential Equations
by C. M. Chen and T. M. Shih

Vol. 10 Statistical Quality Control — A Loss Minimization Approach
by D. Trietsch

Vol. 11 The Mathematical Theory of Nonblocking Switching Networks
by F. K. Hwang

Vol. 12 Combinatorial Group Testing and Its Applications (2nd Edition)
by D.-Z. Du and F. K. Hwang

Vol. 13 Inverse Problems for Electrical Networks
by E. B. Curtis and J. A. Morrow

Vol. 14 Combinatorial and Global Optimization
eds. P. M. Pardalos, A. Migdalas and R. E. Burkard

Vol. 15 The Mathematical Theory of Nonblocking Switching Networks (2nd Edition)
by F. K. Hwang

Vol. 16 Ordinary Differential Equations with Applications
by S. B. Hsu

Vol. 17 Block Designs: Analysis, Combinatorics and Applications
by D. Raghavarao and L. V. Padgett

Vol. 18 Pooling Designs and Nonadaptive Group Testing — Important Tools for DNA Sequencing
by D.-Z. Du and F. K. Hwang

Series on
Applied Mathematics
Volume 18

POOLING DESIGNS AND NONADAPTIVE GROUP TESTING

Important Tools for DNA Sequencing

Ding-Zhu Du
University of Texas at Dallas, USA
and
Xi'an Jiaotong University, China

Frank K Hwang
National Chiao Tung University, Taiwan, ROC

NEW JERSEY • LONDON • SINGAPORE • BEIJING • SHANGHAI • HONG KONG • TAIPEI • CHENNAI

Published by

World Scientific Publishing Co. Pte. Ltd.
5 Toh Tuck Link, Singapore 596224
USA office: 27 Warren Street, Suite 401-402, Hackensack, NJ 07601
UK office: 57 Shelton Street, Covent Garden, London WC2H 9HE

British Library Cataloguing-in-Publication Data
A catalogue record for this book is available from the British Library.

POOLING DESIGNS AND NONADAPTIVE GROUP TESTING
Important Tools for DNA Sequencing
Series on Applied Mathematics — Vol. 18

ISBN-13 978-981-256-822-9
ISBN-10 981-256-822-0

Printed in Singapore

Preface

Group testing has been well-known for its various applications in blood testing, chemical leakage testing, electric shortage testing, coding, multi-access channel communication ..., among others, which are well-documented in our book *Combinatorial Group Testing and Its Applications* (1993, 2nd ed. 2000). However, its application into molecular biology, especially in the design of screening experiment, is quite recent and still under tremendous development (a group testing algorithm for this application is often referred to as *pooling design*). For example, in our 1993 book, we mentioned this application only on one page. In the 2000 edition we covered this topic in three chapters. But now we have to write a whole book to report its progress, and to make painful choices of what to include.

The new application also brings new problems to the theory of group testing. We mention the three fundamental differences.

1. To achieve a given screening objective, the number of biological experiments required is huge and each such experiment is time-consuming (compared to electronic testing). Thus it is of utmost importance to use parallel, or nonadaptive, testing where all experiments can be performed parallelly, or at least in a few rounds. Traditionally, the theory of group testing has been focused on sequential testing, due to its requirement of fewer number of tests. But the focus of pooling design is now shifted to nonadaptive kind. Since a nonadaptive design is just a binary matrix, many mathematical tools can now be brought to group testing for judicious use.

2. A biological experiment is known for its unreliability, as versus electronic testing which carries high accuracy. Therefore error-detection and error-correcting, long ignored in the traditional group testing theory, have to be dealt with squarely. A very lucky and unexpected turn-out is that not only the newly developed theory of pooling design can deal with errors, but in a highly structured way in the sense that treatment of errors is not of the ad-hoc type and all such treatments in different models have the same features.

3. Attempts have been made before to set up traditional group testing in graph models, but such attempts are superfluous in the sense that only terminology, but no

substance, is graph-theoretic. In the complex model of pooling design, we now have a genuine extension of group testing to graph testing from a real need in molecular biology. Not only the theory of group testing is enriched, the extension also provides many new interesting and challenging problems to graph theory.

A one-semester graduate course on the first nine chapters of this book was taught at the National Chiaotung University at Hsinchu in Spring 2005. Chapters 6 and 7 were taught as a 3-week short course at the National Center of Theoretical Sciences. The second author thanks the institutions for financial support and the students for participation. We wish to thank Dr. Hong Zhao, Dr. Ying Liu and Dr. My Thai for their help during preparation of this book.

Contents

1 Introduction

We introduce the notion of group testing with its many industrial applications, and its recent application to various molecular biology screening designs, usually referred to as "pooling designs". We also summarize several mathematical topics which play an important role in developing the theory of pooling designs.

1.1 Group Testing

The theory of group testing has been well developed since 1943. Its combinatorial branch, the so-called *combinatorial group testing*, has been flourishing (see [8] for a general reference) due to its many applications to blood testing, chemical leak testing, electric shorting detection, codes, multi-access channel communication and AIDS screening. Recently, it has also been found to be useful in molecular biology. However, as true in each previous application, this new application also raises new models and new problems such that the classical theory needs to be modified, generalized or simply renovated to handle them. This book is an attempt to contribute to the theory of group testing oriented towards its application to molecular biology.

We first give a brief description of the basic model of combinatorial group testing. There are n items each can be either *positive* (used to be called *defective*) or *negative* (used to be called *good*), while the number of positive items is upper bounded by an integer d. The problem is to identify all positive items. The tool of identification is the so-called *group tests*, while a group test is applicable to an arbitrary subset of items with two possible outcomes; a *negative outcome* indicates that all items in the subset are negative; a *positive outcome* indicates otherwise. The goal is to minimize the number of such tests in identifying the positive items. We will refer to this model as the $(n, \bar{d})$ model while the (n, d) model specifies that d is the exact number of positive items. It was proved [11] that the minimum number of required tests differ by at most 1 between these two models, justifying the frequent use of the (n, d) model due to its mathematical simplicity.

There are two general types of group testing algorithms, sequential and nonadaptive. A *sequential* algorithm conducts the tests one by one and allow a later test to use the outcomes of all previous tests. A *nonadaptive* algorithm specifies all tests simultaneously, thus forbidding using the outcome information of one test to design another test. Sequential algorithms require fewer number of tests in general, since extra information allows for more efficient test designs. Nonadaptive algorithms permit to conduct all tests simultaneously, thus saving the time for testing if not the number of tests. Between the sequential and the nonadaptive algorithms, there are the s-stage algorithms for which all tests in a stage must be specified simultaneously, but the stages are sequential. In particular, an s-stage algorithm is said to be *trivial* if in the last stage every test contains only an individual item.

Historically, the main goal of group testing is to minimize the number of tests. Therefore sequential algorithms have dominated the literature. Note that for the (n, d) model, the information-theoretical lower bound (heretofore referred to as the *information bound*) is

$$\lceil \log_2 \binom{n}{d} \rceil \sim d \log_2(n/d).$$

Since we can use the binary splitting algorithm to identify a positive item in at most $\lceil \log_2 n \rceil$ tests, $d\lceil \log_2 n \rceil$ tests suffice for the (n, d) model. By the closeness of $d\lceil \log_2 n \rceil$ with the information bound, we can say the (n, d) model is practically solved. On the other hand, the determination of exact optimal (n, d) algorithm is very difficult. Let $t(n, d)$ denote the minimum number of tests. $t(n, 1)$ is known to match the information bound, but $t(n, d)$ is not completely determined for any $d \geq 2$.

In the application to molecular biology, a group testing algorithm is called a *pooling design* and the composition of each test a *pool*. While it is still important to minimize the number of tests, two other goals emerge. In molecular biology, a group test corresponds to a PCR experiment (to be explained later) which could take several hours, where the items corresponding to the set of molecules under study could be in the tens of thousands. In some biological problem we have to identify different types (in thousands) of defectives while each type requires a separate group testing procedure, thus performing the tests sequentially is impractical though it requires fewer tests in general. The focus then is *nonadaptive group testing algorithms*, in which all tests are performed simultaneously (thus the information on the outcome of one test cannot benefit the selection of items in another test), although sometimes a pooling design has to settle for 2-stage or s-stage design for small s. Occasionally, sequential procedures can still be used, but the total time needed to identify the positive items must be considered along with the total number of tests.

Biological experiments are known to be unreliable. So the second goal is to control the experimental error which has seldom been studied in the classical group testing literature. This can be done in two ways: The first is to build in error-tolerance in the pooling design so that even though errors occur, the positive items can still be correctly identified; the second is the ability to estimate the effect of errors and the probability of their occurrence.

It is harder to construct nonadaptive group testing algorithms than sequential ones. For given n, often we do not know whether there exists a pooling design not exceeding t pools (tests). For that reason random pooling designs have been proposed which exist for all n and all t. Although random pooling designs are surprisingly efficient, they do not guarantee the identification of all positive items. Thus it is important to be able to estimate the probability of a given positive item not being identified, and the probability of all positive items being identified.

Group testing has been extended to graph testing. A graph testing model has three parameters $(G, I(G^*), P)$ where G is a given graph with vertex-set $V(G)$ and egde-set $E(G)$, $I(G^*)$ is what we know about the hidden subgraph G^* to be identified, and P is the property of a test on a subgraph that is required to have a positive outcome. (If $I(G^*) = \emptyset$, then we can abbreviate to (G, P).) Such a problem is also known as "*learning a hidden subgraph*". Typically, a test on a subset G' gives a positive outcome if and only if G' contains P. For example, the (n, d) group testing model can also be interpreted as a (K_n, $|G^*|$ is a set of d positive vertices, any positive vertex) model, where K_n is the complete graph with n vertices. We may call it the *vertex-testing* model. Of course, this graph representation of the group testing model is superfluous since we only deal with the vertices in K_n and in G'.

The second special case of graph testing is *edge-testing*. A typical model is $(G, I(G^*), \text{any edge in } E(G^*))$. Since the P part is always the same in edge-testing, we will omit it to save space. In particular, if the only information about G^* is that it has d (or at most d) edges, we write (G, d) (or $(G, \bar{d})$ or simply (G) if d is unknown). Such a model was first studied by Chang and Hwang [5] in the special case $(K_{m,n}, 1)$, where $K_{m,n}$ is the complete bipartite graph with m and n vertices, respectively, and an edge e is the unknown positive. They proved that $t(K_{m,n}, 1) = \lceil \log_2 n \rceil$ (the information bound). Aigner [2] generalized $K_{m,n}$ to G and conjectured that $t(G, 1)$ differs from the information bound $\log_2 |E(G)|$ only by a constant. Du and Hwang [8] further conjectured that the constant is 1, which was proved by Damaschke [7].

We can generalize the above two models to the $(G, \text{a } d\text{-set of } K_k, K_k \in G^*)$ model (the above two models correspond to $k = 1, 2$). For better representation, the hypergraph H is used to replace G. Then G^* is a d-set of $E(H)$. When P is any positive edge, we represent this model by $(H : d)$. Note that the hypergraph can also accommodate the case that the hyperedges do not have the same number of vertices. Triesch [21] extended Damaschke's result to the $(H : 1)$ model.

Surprisingly, the (n, d) group testing model can also be represented as a $(K_n^d, 1)$ edge-testing model where K_n^d is the *complete hypergraph* with n vertices and $\binom{n}{d}$ edges each having d vertices. To see this, note that a test on a subset $S \subseteq V$ in the (n, d) model has two outcomes: either S contains no vertex in K_d, or the opposite. On the other hand, a test on $\bar{S}$ (the complementary set of S) in the $(K_n^d, 1)$ model also has two outcomes: either $\bar{S}$ contains the hyperedge e, which is equivalent to K_d, or the opposite. But, $\bar{S}$ contains e if and only if S contains no vertex of K_d. So, testing S in the first model is equivalent to testing $\bar{S}$ in the second model. Note that this conversion from G to H does not work if P is not K_d, for example, P is a star, since a subset (a hyperedge) does not reveal the center of a star.

The $(H : d)$ model represents a total breaking from the group testing model since the correspondence between group testing and graph testing is only for $(H : 1)$. Johann [12] proved a conjecture of Du and Hwang that $t(G, d)$ differs from the information bound $d \log_2 \lceil |E(G)|/d \rceil$ only by dc. Recently, Chen and Hwang [6] further extended this result to hypergraphs. In particular, they did it without assuming the knowledge of d as all other above mentioned results do.

For the edge-testing model, Grebinski and Kucherov [9] first studied the case when there is a structural information about G^*. They consider the case that G^* is a Hamiltonian cycle. Later, Alon, Beizel, Kasif, Rudth and Sudarov [1], and Hwang and Lin [10] considered the case that G^* is a complete matching. Grebinski and Kucherov also considered the more general case when G^* is a graph of maximum degree Δ.

1.2 Nonadaptive Group Testing

A nonadaptive group testing scheme can be represented as a 0-1 (or binary) matrix $M = (m_{ij})$ where columns are labeled by items and rows by tests. Thus m_{ij} specifies that test i contains item j. Sometimes it is more convenient to view a 0-1 column C_j as the incidence vector of subset $\{i \mid m_{ij} = 1\}$. Then we can talk about the union $U(s)$ of a set s of columns, which is nothing but the boolean sum of the corresponding 0-1 columns. Similarly, we can view a row as the incidence vector of a subset of the column indices where its 1-entries lie.

Given a set D of positive items, the outcomes of the tests can also be represented as a 0-1 vector where 0 indicates a negative outcome, i.e., all items in the test are negative, and 1 indicates a positive outcome, i.e., the test set contains at least one positive item. Note that the outcome vector can be represented by $U(D)$, the union of the columns in D.

A minimum requirement for M to be able to identify D is that $U(D) \neq U(D')$ for $D \neq D'$. A matrix with this property is called *$\bar{d}$-separable* if $|D| \leq d$ is assumed, and *d-separable* if $|D| = d$ is assumed. Although in theory, the vector $U(D)$ uniquely determines D, the actual decoding can be messy. Thus one can trade off weaker requirement for an easy decoding. M is called *d-disjunct* if no column is contained in the union of any other d columns. Then the decoding for a column C is $C \in D$ if $C \subseteq U(D)$. For convenience, we use *d-separating* as a generic term for d-separable, $\bar{d}$-separable or d-disjunct.

Let M be a d-separating matrix. Suppose a row R is contained in another row R'. Then we can replace R' by $R' \setminus R$ without losing any information (and may gain some) since R being negative implies testing R' is the same as testing $R' \setminus R$, where R being positive implies testing R' provides no further information. Therefore, we may assume throughout the book that M contains no row properly contained by another row unless specified otherwise. Consequently, if a row R contains a single 1-entry, then the column C incident to R contains no other 1-entry. We call R an isolated row and C an isolated column. If such a pair exists, we learn the state of C immediately, and can discard C and R from M since they have no effect on the testing of other columns except d needs to be replaced by $d-1$ if C is positive. We also often assume that M contains no isolated row or column.

We give examples of d-separating matrices with $d = 1$.

1-separable. All we need is that columns are distinct. Let B_t denote the binary matrix consisting of all 2^t distinct t-dimensional vectors. Then for $n \leq 2^t$, any n columns constitute a 1-separable matrix for n items. The decoding is trivial, i.e., $C \in D$ if $C = V(D)$. For example, B_3 is

$$\begin{pmatrix} 0 & 0 & 0 & 0 & 1 & 1 & 1 & 1 \\ 0 & 0 & 1 & 1 & 0 & 0 & 1 & 1 \\ 0 & 1 & 0 & 1 & 0 & 1 & 0 & 1 \end{pmatrix}$$

The *weight* of a column is the number of 1's in it. Suppose there is an upper bound $\bar{w}$ of column weight. Consider the submatrix B_t^w consisting of all columns with weight $\leq \bar{w}$, there are

$$\sum_{i=0}^{\bar{w}} \binom{t}{i}$$

columns. For n not exceeding this number, any n columns of B_t^w constitute a 1-separable matrix satisfying $\bar{w}$.

Suppose there is an upper bound k of pool size. We demonstrate a construction of a special case $n = g^{x+1}$ and $k = g^x$. The tests are divided into groups where each group has $g-1$ tests. In the first group the n items are partitioned into disjoint k-subsets, and each k-subset except the last constitutes a test. In the second group, the n items are re-partitioned into disjoint k-subsets such that each pair of k-subsets, one from the first group and one from the second group, intersect in exactly g^{x-1} items. Again, test every k-subset except the last. In general, in the ith group, $2 \leq i \leq x$, we require each k-subset to intersect every k-subset in previous groups in exactly g^{x-1} items. The resulting matrix is 1-separable with $(g-1)(x+1)$ tests of size k. For example, for $n = 27$ and $k = 9$, the matrix is

$$\left(\begin{array}{ccccccccccccccccccccccccccc}
1&1&1&1&1&1&1&1&1&&&&&&&&&&&&&&&&&& \\
&&&&&&&&&1&1&1&1&1&1&1&1&1&&&&&&&&& \\
1&1&1&&&&&&&1&1&1&&&&&&&1&1&1&&&&&& \\
&&&1&1&1&&&&&&&1&1&1&&&&&&&1&1&1&&& \\
1&&&1&&&1&&&1&&&1&&&1&&&1&&&1&&&1&& \\
&1&&&1&&&1&&&1&&&1&&&1&&&1&&&1&&&1&
\end{array}\right)$$

For general n and k, the construction principle is the same except the partition would not be so balanced. In the formula for the number of tests, g will be replaced by $\lceil n/k \rceil$ and x by

$$\lceil \log_{\lceil n/k \rceil} n \rceil = \lceil \frac{\log n}{\log \lceil n/k \rceil} \rceil.$$

$\bar{1}$**-separable**. The only difference from 1-separable is that the 0-vector must be deleted from M (to be reserved for the outcome $D = \emptyset$).

1-disjunct. Spencer [19] proved that for given t of tests, the maximum number n of items in a 1-disjunct matrix is

$$\binom{t}{\lfloor t/2 \rfloor},$$

i.e., the matrix consists of all $\lfloor t/2 \rfloor$-subsets of the set $\{1, ..., t\}$. For n not exceeding that number, any n columns constitute a 1-disjunct matrix for n items. For $t = 5$, the maximum matrix is

$$\begin{pmatrix} 1 & 1 & 1 & 1 & 0 & 0 & 0 & 0 & 0 & 0 \\ 1 & 0 & 0 & 0 & 1 & 1 & 1 & 0 & 0 & 0 \\ 0 & 1 & 0 & 0 & 1 & 0 & 0 & 1 & 1 & 0 \\ 0 & 0 & 1 & 0 & 0 & 1 & 0 & 1 & 0 & 1 \\ 0 & 0 & 0 & 1 & 0 & 0 & 1 & 0 & 1 & 1 \end{pmatrix}$$

Suppose $\bar{w}$ is an upper bound of column weight. Schultz, Parnes and Srinavasan [17] proved that the maximum number of columns is $\binom{t}{\bar{w}}$.

Suppose k is an upper bound of test size. Note that $k = \binom{t-1}{\lfloor t/2 \rfloor - 1}$ for each row in Spencer's result. Therefore for $k \geq \binom{t-1}{\lfloor t/2 \rfloor - 1}$, Spencer's result remains to be the best construction. For $n > k^2/2$, Cai [3] constructed a 1-disjunct matrix with $\lceil 2n/k \rceil$ rows obtained by the incidence matrix of a graph with $\lceil 2n/k \rceil$ vertices, n edges and every vertex except possibly one having degree k (it is easy to construct such a graph). For example, for $n = 12$ and $k = 3$, we have the graph in Fig. 1.1 and the matrix as

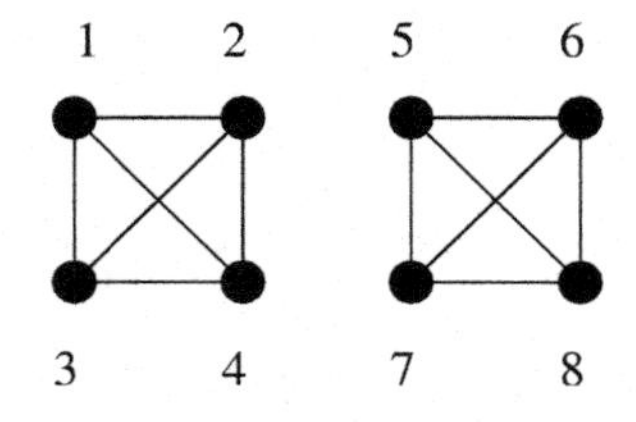

Figure 1.1: A 3-regular graph.

follows:

	12	13	14	23	24	34	56	57	58	67	68	78
1	1	1	1									
2	1			1	1							
3		1		1		1						
4			1		1	1						
5							1	1	1			
6							1			1	1	
7								1		1		1
8									1		1	1

For $k^2/2 \geq n \geq \binom{k}{2}$, Ramsey and Roberts [16] proved $t(d, n, k) = k + 1$. Finally, for $\binom{k}{2} > n \geq 2k$, Knudgen, Mubayi and Tetali [13] proved

$$\min\{w \mid n \leq \binom{w}{\lfloor wk/n \rfloor}\} \geq t(d, n, k) \geq \min\{w \mid n \leq \binom{w}{\lceil wk/n \rceil}\}.$$

1.3 Applications in Molecular Biology

An eukaryotic cell has a nucleus. The nucleus contains chromosomes which carry the genetic instructions for making living organism. Cells in different kinds of organisms contain different numbers of chromosomes. Each human cell contains 23 pairs of chromosomes. Each chromosome is a packed DNA and is 5,000 times shorter than the extended form of DNA (Fig. 1.2).

Each DNA consists of two chains, entwined forming a double spiral to form a DNA double helix. Each chain is a sequence of four types of nucleotides, A, C, G, T, and hence can be seen as a string of alphabets $\{A, C, G, T\}$. The two strings stay together in double helix with a duality relation where one string can be obtained from the other by the mapping $A \to T$, $C \to G$, $G \to C$, $T \to A$. The two strings can be separated under proper heating, but they have a tendency to bind to each other when put together. This binding tendency is known as *hybridization.*

We can use hybridization to find out whether a DNA string T, called a *target sequence*, contains a specific substring S. Let S^{-1} denote the complementary strand of S. Mix T and S^{-1}. If T contains S, then S^{-1} will hybridize with S. This hybridization effect is magnified by the PCR (polymer chain reaction) technique so that it becomes observable or measurable.

A PCR is essentially a technology to make copies of existing DNA pieces by using hybridization. A *primer* is a short DNA sequence either preceding or succeeding the DNA sequence S (S^{-1}) of interest. If S exists in T, the hybrization effect would make a primer of a S^{-1} to grow into a S^{-1} along S under action of enzyme. Heating is able to break up the hybridized pair to allow each half to grow with other copies of S and S^{-1}. This (break-up, hybridize) cycle reiterates until a huge PCR product is obtained (Fig. 1.3). Hence the observation of a PCR product when primers of S^{-1}

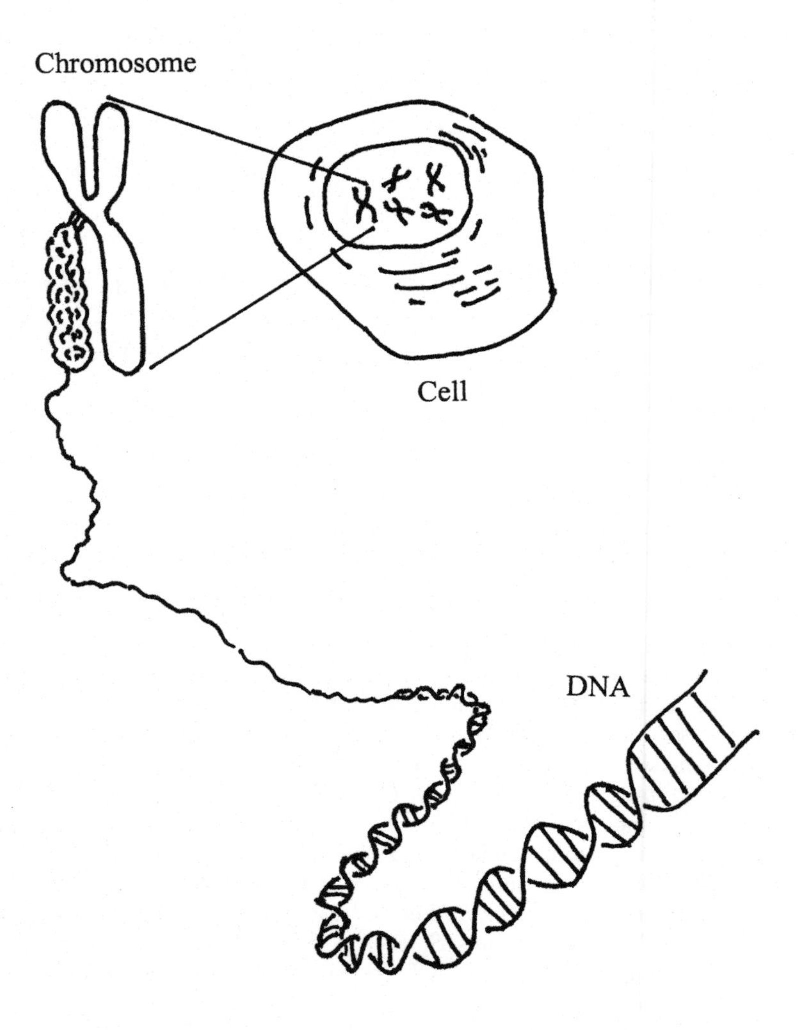

Figure 1.2: Each chromosome is a packed DNA.

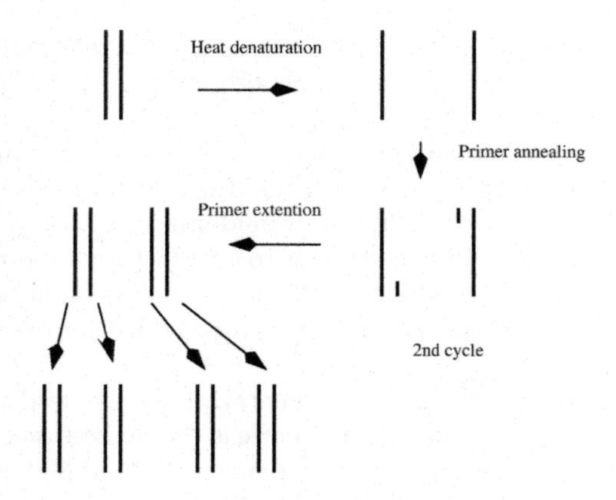

Figure 1.3: Polymer Chain Resaction (PCR).

are used as the *probe* implies the existence of S in T. In fact, suppose we mix T with the two primers sandwiching S^{-1} and S is not too long. Then a PCR product will grow whose length corresponds to the length of S. This length factor is critically used in the multiplex PCR.

In a multiplex PCR, many primers are used simultaneously in one test. Suppose the set P of primers are ordered according to their positions in T (we do not know the order). Then two primers are *adjacent* if there is no other primer in P between them and they are not too far apart in T. In a multiplex test, a PCR product is obtained for each pair of adjacent primers. In theory, by counting the number of PCR products of different lengths, we know the number of pairs of adjacent primers. In practice, this information is not entirely accurate due to experimental errors and also due to the possibilities that two PCR products have similar lengths.

For our purpose we classify PCR into two categories. The *duality PCR*, using S^{-1} to catch S, basically tests for the containment relation. The other, the *primer* PCR, tests the adjacency relation between primers. Using the multiplex PCR we have a choice of whether to use the regular group testing model with only "yes" or "no" outcome, or the quantitative model which specifies the number of adjacent pairs of primers.

We now mention some specific applications of pooling designs in molecular biology.

(i) Physical mapping. A clone library stores the DNA sequence T of a large molecule, such as a chromosome. Typically, the large molecule has to be broken into manageable sizes for easy storage, multiplication and study. The broken pieces are called *clones*. The cutting can be done either by a restriction enzyme which cuts whenever a given short DNA sequence, about six nucleotides and varying with enzymes, is encountered, or a random cutting called *shotgunning* which is effected by applying heat or shock. Regardless the cutting methods, the important things are:

(a) one cannot control lengths of the cuts, (b) one can get different cuttings through using different enzymes or through several shotgunnings, (c) the order of the clones is lost through a cutting.

When we want to study the target sequence in its entirety, we have to reconstruct the order of clones. This is a problem since even if we could read the clones, which we can't if the length of a clone is over 700, we would still be at a loss as to which clone should follow which. The prevailing method is to cut the target sequence several times into clones, each time with a different restriction enzyme or shotgunning to obtain different cuttings. Then we use the information provided by overlapping clones to reconstruct the target sequence.

As we said, we cannot read a clone in general. Thus we need to identify certain characteristics of a clone. One such characteristic is the possession of STSs (sequence-tagged sites) where each STS is a short DNA sequence (about 200 nucleotides), which appears uniquely in the target sequence. For each clone we identify the subset of STSs it contains. If two clones share an STS, then they must be adjacent in the target sequence. Note that when the clones are sequenced, the set of STSs marking the clones also becomes a sequence marking the target sequence, which is usually referred to as a *physical mapping* of the target sequence. Figure 1.4 illustrates the idea of using STSs to sequence the clones.

target sequence	A		B		C		D		E		F	G
first cut	A		B	|	C		D	|	E		F	G
second cut	A	|	B		C	|	D		|		F	G
third cut	A		B		|		D		E	|	F	G

overlapping graph

AB -- BC -- CD

|

EFG -- DE

Figure 1.4: Using STSs to sequence.

Suppose the target sequence has seven STSs: A, B, C, D, E, F, G. There are three cuttings resulting in a total of ten clones. Note that the second cutting cuts right into E, so E does not appear in any clone in the second cutting. Delete clones whose STS-set is a subset of another clone, and retain only one among those having the same STS-set. Draw an overlapping graph with the surviving clones as vertices and an edge between two clones if there is an overlap. A Hamiltonian path of the surviving clones then suggests a possible target sequence. Note that there may exist several Hamiltonian paths or none, and one Hamiltonian path may suggest several target sequences.

Before this, we need to identify the STS-set for each clone. We do this by one STS at a time. Suppose we are doing A. Then we use A^{-1} as the probe in a pooling design in which a clone is positive if it contains A, and negative if not. Note that the number of positive clones cannot exceed the number of cuttings (assuming no error), but can be less since an STS may be separated in a cutting or contaminated. Therefore we can set d to be the number of cuttings. Also note that thousands of probes must be used to obtain enough information on the clones. Thus sequential procedures are impractical.

(ii) Contig sequencing

After we use the overlapping information to sequence the clones, it is still possible no Hamiltonian path exists. Typically, the clones are sequenced into a set of relatively long molecules, called *contigs*, separated by gaps. For example, in Fig. 1.4, if the first cutting also cuts into E, then we will have two contigs, one consisting of $ABCDE$ and the other FG, separated by the gap from the end of E to the start of F. Multiplex PCR can be used to sequence the set of contigs. Suppose there are n contigs. The set of primers consists of the two end-subsequences (about 20 nucleotides long) of each contig. We call two contigs, or two primers, *adjacent* if they are separated by a single gap.

For a pooling test we mix a subset of primers with the target sequence. If the mixture contains a pair of adjacent primers, then a PCR product will be produced whose length corresponds to the length of the gap between them. If the mixture contains several such pairs, then each pair yields a PCR product, and we can distinguish them by their lengths in theory. This problem can be translated into an edge testing problem where each primer is a vertex, and each pair of adjacent primers is a positive edge (alternatively, one can also treat each contig as a vertex, and each pair of adjacent contigs a positive edge), while the underlying graph G is the complete graph. However, we do have the extra information that the subgraph consisting of the positive edges is a Hamiltonian path (when contigs are the vertices) or a perfect matching (when primers are the vertices).

Since the number of contigs is usually not too large and only one pooling is needed, sequential procedures can be considered for the contig sequencing problem.

(iii) Locating STS among ordered clones

Suppose the clones are ordered and we want to identify all clones containing a given STS. Due to the uniqueness of an STS in the tag sequence, clones containing the STS must be consecutively ordered, and their number is bounded by the number of cuttings generating these clones. This is a new group testing problem in which not only the objects are linearly ordered, but also the positive objects are consecutive.

(iv) Finding exon boundaries in cDNA

A human gene is often split into several disjoint parts, called *exons*, separated by gaps, called *introns.* The DNA sequence of a gene, including all introns, is first transcribed to mRNA. Then the introns are edited out before the corresponding proteins are made. We can backtrack this process to obtain the gene with introns edited out from the proteins, called cDNA. The problem is to identify the boundaries between exons and introns, called *exon boundaries*, in cDNA so to discover the composition of the target sequence. Here the existence of an exon boundary is equivalent to the existence of an intron.

A solution to this problem was given in [22]. A suitable number of primers was developed from the cDNA sequence, "suitable" in the sense that the distance between each pair of adjacent primers satisfies the design specification. A probe is a pair of primers applying to both the cDNA interval (between the two primers) and the target sequence. The test outcome is obtained by comparing the lengths of the PCR products from the cDNA interval and the target sequence. If the latter is longer, it must be because the target sequence contains introns though their exact number and locations are unknown, while the cDNA obviously does not. Note that two primers too far apart cannot produce a PCR product. Hence the length of the cDNA interval for a test may be restricted.

Call an interval positive if it contains an intron. Then the problem is to identify all positive intervals, a group testing problem, except tests are restricted to sets of consecutive items. One would like to minimize both the number of rounds of testing and the number of distinct primers, since we need to do n probings, sequential procedure can take too much time.

(v) Protein-Protein Interaction

Protein-protein interactions are critical in many biological processes, such as the formation of macromolecular complexes and the transduction of signals in biological pathways. The interaction is usually between a bait protein and a prey protein. Therefore, to identify all protein-protein interactions, we are facing a group testing problem in bipartite graph with bait proteins as one vertex set and prey proteins as the other vertex set [20]. Interactions are edges between the two vertex sets.

(vi) Non-unique Probe Selection

To identify closely related virus subtypes in a biological sample, it is hard to find unique probes (i.e., each hybirizes only one target). In this situation, Schliep, Torney and Rahmann [18] suggested to use non-unique probes with group testing techniques. In the case that the number of targets in a sample is expected to be small, the number of probes can also be reduced. For example, on a data set of 285 rDNA sequences, they were able "to idetify 660 sequences, a substantial improvement over a prior approach using unique probes which only identified 408 sequences."

(vii) The complex model

Suppose the items are molecules. Then a biological function may depend on the presence of a subset of molecules, called a *complex*. Therefore, the notion of a positive molecule is transformed to a *positive complex*. The complex model can also be easily fitted into the "group testing on hypergraph H" framework. Treating each molecule as a vertex, then a complex is simply an edge in H.

Thus the complex model intensifies the need of studying group testing on hypergraphs. We will see that important breakthroughs have been made on this model which not only meets the real biological need, but also advances the theory of group testing.

Although only one probing is required for this problem and thus sequential procedures can be considered, the number of tests is huge and time-consuming. Nonadaptive or k-round procedures will still be desired.

1.4 Pooling Designs for Two Simple Applications

The applications of group testing to problems (iii) and (iv) in the last section are relatively straightforward, and will be discussed in this section.

First, for the exon boundary problem, Xu *et al.* [22] suggested the following procedure. A typical pooling design for this problem may consist of several rounds of testing. Suppose we estimate the number of inxons to be k. The cDNA sequence is first partitioned into n equally spaced intervals. Primers at both ends of these intervals are developed. A $\bar{d}$-separable matrix is used to identify the positive intervals. In the second round, we further divide each positive interval into sub-intervals to reduce the length of a positive interval. If the positive subintervals identified in the second round are still considered too long, further divisions are taken. When to stop depends on a balance of several conflicting goals: small positive intervals, small number of rounds, small number of tests and small number of primers (the same primer can be used in multi-rounds). Note that the multi-round structure is a balance between a sequential procedure, which uses too many rounds, and a 1-round procedure, which uses too many tests and primers.

Next comes the problem to identify all up-to-d consecutive positive clones in an ordered set of n clones. Colbourne [4] gave a sequential algorithm with $\log n + \log d + c$ tests where t is a constant. We make a slight improvement by eliminating c.

Use the halving procedure to identify the first positive clone in $\log n$ tests. Suppose it is C_i. Then the only uncertainty is about clones $C_{i+1}, ..., C_{i+d-1}$. Use the halving procedure again, but in a reverse order, to identify the last positive clone, if any, in $\log d$ tests. If it does not exist, then C_i is the only positive clone. If C_{i+j} is the last positive clone, then $\{C_i, ..., C_{i+j}\}$ is the set of positive clones. Colbourne also gave a nonadaptive algorithm. First, we introduce the notion of a Gray code G_n of order n which is a sequence of 2^n binary n-vectors such that two consecutive vectors differ only in one bit. G_1, G_2, G_3 are given below:

$$\begin{array}{ccc} & & 00001111 \\ & 0011 & \\ 01 & & 00111100 \\ & 0110 & \\ & & 01100110 \\ & & \\ G_1 & G_2 & G_3 \end{array}$$

In general, G_n can be obtained from G_{n-1} by taking two copies of G_{n-1}, reversing the second copy and adding a first row which has 0s in the first copy and 1s in the second copy as illustrated below:

		0000000011111111
00001111	11110000	0000111111110000
00111100	00111100	0011110000111100
01100110	01100110	0110011001100110
G_3	G_3^{-1}	G_4

A matrix is called 2-consecutive if it can identify all positive clones provided there are exactly two which are consecutive; it is called $\bar{2}$-consecutive if "exactly two" is replaced by "up to two". We now show how to construct 2-consecutive and $\bar{2}$-consecutive matrices from G_n.

2-Consecutive. Note that for any two consecutive columns, one is a subset of the other. Therefore the union of two consecutive columns always equals to the larger column. With the outcome vector V, we first identify column $C_i = V$. For $d = 2$, the other positive column is either C_{i-1} or C_{i+1}. So it suffices to add two rows such that $C_{i-1} \neq C_{i+1}$ in those two rows, and C_i does not contain both C_{i-1} and C_{i+1}. One way is to let the ith such row, $i = 1, 2$, has 1s in column j if and only if $j \equiv i \pmod 4$. The following is such a matrix for eight items.

$$\begin{matrix} 0&0&0&0&1&1&1&1\\ 0&0&1&1&1&1&0&0\\ 0&1&1&0&0&1&1&0\\ 1&0&0&0&1&0&0&1\\ 0&1&0&0&0&1&0&0 \end{matrix}$$

$\bar{2}$-Consecutive. Then C_i itself can be the positive set. We need to add three rows to differentiate the three cases. One way is to let the ith such row, $i = 1, 2$, have 1s in column j if $j \equiv i \pmod 3$. An example for $n = 8$ is

$$\begin{matrix} 0&0&0&0&1&1&1&1\\ 0&0&1&1&1&1&0&0\\ 0&1&1&0&0&1&1&0\\ 1&0&0&1&0&0&1&0\\ 0&1&0&0&1&0&0&1\\ 0&0&1&0&0&1&0&0 \end{matrix}$$

Note that the $D = \emptyset$ case is also taken care of since the 0-vector is not in the matrix.

If $d = 2$, use a $\bar{2}$-consecutive matrix to identify the positive clones. If $d > 2$, then partition the n clones consecutively into groups of $d - 1$ (the last group can have fewer). Call a group positive if it contains a positive clone. Then at most two consecutive groups can be positive. Treat each group as an object and use a $\bar{2}$-consecutive matrix to identify the positive group(s) if any, in at most

$$\left\lceil \log \left(\frac{n}{d-1} \right) \right\rceil + 3 \text{ tests.}$$

Colbourne proposed to add $2(d-1)$ rows to identify which clones in the positive group(s) are positive. But actually, adding $d+1$ rows suffices where row i, $1 \le i \le d+1$, has 1 in column j if $j \equiv i \pmod{d+1}$.

Let $n(d_{cons}, t)$ denote the maximum n such that there exists a $t \times n$ d-consecutive matrix. Similarly, $n(\bar{d}_{cons}, t)$ is defined. Since in the $\bar{2}$-consecutive problem, n columns generate $2n-1$ sample points whose outcome vector must be all distinct in the $(\bar{2}_{cons}, t)$ space. Hence $n \le 2^{t-1}$. Müller and Jimbo [15] showed that this upper bound can always be achieved except for $t=3$ by a recursive construction. Let $M_t = (C_1, C_2, ..., C_n)$, $n = 2^{t-1}$, denote such a matrix of order t. Then M_2, M_4, M_5 are given explicitly, while for $t \ge 4$,

$$M_{t+2} = \begin{pmatrix} 0 & 0 & C_n & \cdots & C_1 & C_2 & \cdots & C_{n-1} & C_n & C_{n-1} & \cdots & C_2 & C_1 & C_2 & \cdots & C_n \\ 0 & 1 & 0 & & 0 & 1 & & 1 & 1 & 1 & & 1 & 0 & 0 & & 0 \\ 1 & 0 & 1 & & 1 & 0 & & 0 & 0 & 1 & & 1 & 0 & 0 & & 0 \end{pmatrix}.$$

They also showed similar constructions work under constant column weight.

1.5 Pooling Designs and Mathematics

Group testing has been traditionally associated with combinatorics and probability as the names *combinatorial group testing* and *probabilistic group testing* suggested. Its relation with *algorithm theory* is also obvious.

A pooling design is a combinatorial design whose blocks are the pools and whose items are the clones (or molecules), except that the requirements are different from the usual *balanced appearances* type. Nevertheless, the whole spectrum of combinatorial designs has lent its strength to the construction of pooling designs. With the close relation to combinatorial designs comes the unavoidable relation to coding theory. It must be said that the relation is not just to put known results in combinatorial design theory and coding theory in good use, but also that pooling designs suggest a new type of designs and also a new type of codes know as "superimposed codes".

As combinatorial designs draw heavily from algebraic and geometric structures, so do pooling designs. Mathematical structures are always helpful in understanding, motivating and finding pooling designs. For example, look at again the construction of 1-separable matrix for $n = g^{x+1}$ items and group size $k = g^x$ in Section 1.2. From the previous description, it is not so easy to see why the construction works, where the idea comes from and how to do specifically. Now, let us represent each item by a $(x+1)$-dimensional vector with all components chosen from g elements $0, 1, ..., g-1$. Then g^{x+1} items would form a $(x+1)$-dimensional cube. For each coordinate, there are g parallel hyperplanes, perpendicular to it, which partition all items into g subsets of k items. Performing tests on $g-1$ of them would reveal which hyperplane contains the unique positive item. So, these k items form a group in the previous description. Clearly, two non-parallel hyperplanes intersect with g^{x-1} items and the positive item would be identified when all coordinates are determined. It can also be easily seen that if all g subsets in each group are tested, then these tests correspond to a $\bar{1}$-separable matrix.

We will find that finite fields, linear spaces in finite field, lattices, t-designs and packing provide very helpful tools for pool designs. It is also well-known that the extremal set theory is closely related to nonadaptive group testing, hence pooling designs.

There are two areas which have only played minor role in traditional group testing, but are of utmost importance in pooling designs, and in the meantime offer great challenges. Group testing have been extended to graphs, but no real motivation to study that problem is known. Now we know that the complex model corresponds exactly to group testing on hypergraph, while the contig sequencing problem and the gene detection problem correspond to the case when we know some property of the hidden subgraph. Therefore an urgency to develop a theory of group testing on graph suddenly emerges.

Traditionally, combinatorial group testing has very little to do with probability theory. Since the construction of pooling design is so much harder than sequential group testing algorithms, the importance of random designs is increased. It turns out that the probability analysis of random designs is a difficult problem, and so far exact analysis have been obtained only for some simple models.

To summarize, we see that group testing and pooling designs not only have surprisingly many applications to various fields, but also borrows heavily from many mathematical branches, and in the meantime, propose new problems to these branches. The mutually beneficial relation has just begun to unreel and there is a lot of work to be done.

For convenience of the reader, we collects some basic knowledge about finite fields simplicial complexes and linear spaces over finite fields in the rest of this section.

A finite non-empty set F with two binary operations, addition $+$ and multiplication $\cdot$, is called a *finite field*, if the following hold

(1) F is a commutative group with addition, that is,

(1a) $x + y = y + x$ for all $x, y, z \in F$,

(1b) $x + (y + z) = (x + y) + z$ for all $x, y, z \in F$,

(1c) F contains an element 0 such that for every element x in F, $x+0 = 0+x = x$, and there exists element $-x$ satisfying $x + (-x) = 0$;

(2) F is a commutative semi-group and $F - \{0\}$ is a commutative group with multiplication, that is,

(2a) $x \cdot y = y \cdot x$ for all $x, y, z \in F$,

(1b) $x \cdot (y \cdot z) = (x \cdot y) \cdot z$ for all $x, y, z \in F$,

(1c) $F - \{0\}$ contains an element 1 such that for every element x in $F - \{0\}$, $x \cdot 1 = 1 \cdot x = x$, and there exists element x^{-1} satisfying $x \cdot x^{-1} = 1$;

(3) $x \cdot (y + z) = x \cdot y + x \cdot z$ for all $x, y, z \in F$;

(4) $0 \cdot x = 0$ for all $x \in F$.

The *order* of a finite field is the number of elements in the filed. It is a well-known fact that a finite field of order q exists if and only if q is a prime power. Moreover, under isomorphism, the finite field of order q is unique for each prime power q, denoted by $GF(q)$ and called the *Galois field* of order q.

When q is a prime, $GF(q)$ consists of all residues modulo q, that is, all elements in Z_q. When $q = p^n$ for a prime p and $r \geq 2$, irreducible polynomial $p(x)$ of degree n.

In fact, for any positive integer n, there exists an irreducible polynomial $p(x)$ of degree n over $GF(p)$, such that the set $\{x, x^2, ..., x^{p^n-2}, x^{p^n-1} = 1\}$ modulo $p(x)$ is the set of all non-zero polynomials of degree less than n, which is also the set of non-zero elements of $GF(p^n)$. In other words, the multiplication group of $GF(p^n)$ is cyclic. The set of scalars (polynomials of degree zero) is a cyclic subgroup of order $p-1$, in the form $\{x^m, x^{2m}, ..., x^{(p-1)m}\}$ where $m = \frac{p^n-1}{p-1}$. We call such $p(x)$ a *primitive polynomial.*

Denote $[t] = \{1, 2, ..., t\}$ and a subset of k elements is called a *k-subset.* A $t-(v, k, \lambda)$ design is a family $\mathcal{F}$ of k-subsets of $[v]$ such that every t-subset is contained in at most λ members, called *block* of $\mathcal{F}$. It is easily verified that

$$b = |\mathcal{F}| = \frac{\binom{v}{t}\lambda}{\binom{k}{t}} \tag{1.5.1}$$

and each element of $[v]$ appears in

$$r = \frac{\binom{v-1}{t-1}\lambda}{\binom{k-1}{t-1}} \tag{1.5.2}$$

blocks. Also, it is easily verified that a t-design is a t'-design for $t' < t$.

A $2-(v, k, \lambda)$ design is often referred to as a (v, b, r, k, λ)-*block design* in the literature with b and r as given above. A $t-(v, k, 1)$ design is called a *Steiner* (t, v, k) *packing.*

A $t-(v, k, \lambda)$ design becomes a $t-(v, k, \lambda)$ *packing* if the condition that every t-subset appears in exactly λ blocks is replaced by at most λ blocks.

When we use the $t-(v, k, \lambda)$ design or packing in the construction of pooling design, we often have to replace t by T to avoid a conflict with using t as the number of rows in the test matrix.

1.6 An Outline of the Book

We will first introduce deterministic constructions of pooling designs. Chapter 2 reviews and updates the theory of nonadaptive group testing, which guarantee to identify all positive items if that number does not exceed d, and their error-tolerant version. Chapter 3 reviews some traditional deterministic construction and Chapter 4 introduces the new construction methods, the partial order method with algebraic structure, and the simplicial complex method with geometric structure.

Then we discuss the more practical random designs. An important criterion here is to evaluate various probabilities that items are not correctly identified. Chapter 5 reviews some basic random designs whose probabilities of nonidentification are all solved. The emphasis here is to obtain the formulas which can be computed fastest, a real concern due to the large number of items. The problem of determining the design parameters to minimize the nonidentification probabilities is still open in general. Chapter 5 also studies pooling designs obtained by mixing a random matrix with a deterministic matrix. It even includes totally deterministic matrices, but used beyond their design specifications; hence identification of positive clones can be stated only in probability terms.

The next two chapters each covers an application mentioned in Section 1.3. Chapter 6 studies the complex model, Chapter 7 studies the contig-sequencing problem. The general technique is to use the graph-testing model.

Chapter 8 presents a generalization of pooling design where an item can be positive, negative or inhibitive, meaning its presence in a pool preempts a positive outcome. The problem is still to identify the positive items. Several multi-stage pooling designs have been given in the literature. We focus on the recently found (one-stage) pooling design, whose construction, interestingly, uses the same mathematical tools as the regular pooling design.

Chapter 9 studies the array design actually used in some clone libraries. This design also leads to some new theoretical questions in the graph-design theory.

In Chapter 10, we consider problems related to non-unique probe selection, including computational complexity issues on pooling designs.

References

[1] N. Alon, R. Beigel, S. Kasif, S. Rudth and B. Sudarov, Learning a hidden matching.

[2] M. Aigner, Search problems in graphs, *Disc. Appl. Math.*, 14 (1986) 215-230.

[3] M. C. Cai, On the problem of Katona on minimal completely separating systems with restrictions, *Disc. Math.*, 48 (1984) 121-123.

[4] C. Colbourn, Group testing for consecutive positives, *Ann. Combinatorics*, 3 (1999) 37-41.

[5] G. J. Chang and F. K. Hwang, A group testing problem, *SIAM J. Alg. Disc. Meth.*, 1 (1980) 21-24.

[6] T. Chen and F. K. Hwang, A competitive algorithm in searching for many edges in a hypergraph, 2003, preprint.

[7] P. Damaschke, A tight upper bound for group testing in graphs, *Disc. Math.*, 48 (1994) 101-109.

[8] D.-Z. Du and F.K. Hwang, *Combinatorial Group Testing and Its Applications (2nd edition)*, World Scientific, 1999.

[9] V. Grebinski and G. Kucherov, Optimal reconstruction of graphs under additive model, *Algorithmica*, 28 (2000) 104-124.

[10] F. K. Hwang and W. D. Lin, The incremental group testing model for gap closing in sequencing long molecules, *J. Combin. Opt.*, 7 (2003) 327-337.

[11] F. K. Hwang, T. T. Song and D.-Z. Du, Hypergeometric and generalized hypergeometric group testing, *SIAM J. Alg. Disc. Methods*, 2 (1981) 426-428.

[12] P. Johann, A group testing problem for graphs with several defective edges, *Disc. Appl. Math.*, 117 (2002) 99-108.

[13] A. Knudgen, D. Mubayi and P. Tetali, Minimal completely separating systems of k-sets, *J. Combin. Thy, Series A*, 93 (2001) 192-198.

[14] J. H. van Lint and R. M. Wilson, *A Course in Combinatorics*, Cambridge University Press, Cambridge, 1992.

[15] M. Müller and M. Jimbo, Consecutive positive delectable matrices and group testing for consecutive positives, *Disc. Math.*, 279 (2001) 369-381.

[16] C. Ramsey and I. T. Roberts, Minimal completely separating systems of sets, *Austral. J. Combin.*, 13 (1996) 129-151.

[17] D. J. Schultz, M. Parnes and R. Srinivasan, Further applications of d-complete designs to group testing, *J. Combin. Inform. & Syst. Sci.*, 8 (1993) 31-41.

[18] A. Schliep, D. C. Torney, S. Rahmann, Group testing with DNA chips: generating designs and decoding experiments, *Proceedings of the 2nd IEEE Computer Society Bioinformatics Conference*, 2003.

[19] J. Spencer, Minimal completely separating systems, *J. Combin. Thy.*, 8 (1970) 446-447.

[20] N. Thierry-Mieg, L. Trilling and J.-L. Roch, A novel pooling design for protein-protein interaction mapping, manuscript, 2004.

[21] E. Triesch, A group testing problem for hypergraphs of bounded rank, *Disc. Appl. Math.*, 66 (1996) 185-188.

[22] G. Xu, S.-H. Sze, C.-P. Liu, P.A. Pevzner and N. Arnheim, Gene hunting without sequencing genomic clones: finding exon bounderies in cDNA, *Genomics*, 47 (1998) 171-179.

2
Basic Theory on Separating Matrices

In this chapter, we review and update the basic theory of separating matrices and its error-tolerant versions.

2.1 d-Separable and $\bar{d}$-Separable Matrices

Kautz and Singleton [17] proved the following lemma.

Lemma 2.1.1 *If a matrix is d-separable, then it is k-separable for every* $1 \leq k \leq d < n$.

Proof. Suppose that M is d-separable but not k-separable for some $1 \leq k < d < n$. Namely, there exist two distinct samples s and s' each consisting of k columns such that $U(s) = U(s')$. Let C_x be a column in neither s nor s'. Then

$$C_x \cup U(s) = C_x \cup U(s') \; .$$

Adding a total of $d-k$ such columns C_x to both s and s' yields two distinct samples s_d and s'_d each consisting of d columns such that $U(s_d) = U(s'_d)$. Hence M is not d-separable. If there are only $\ell < d-k$ such columns C_x, then select $d-k-\ell$ pairs of columns (C_y, C_z) such that C_y is in s but not in s' and C_z is in s' but not in s. Then

$$C_z \cup U(s) = U(s) = U(s') = C_y \cup U(s') \; .$$

Therefore these pairs can substitute for the missing C_x. Since M is d-separable, a total of $d-k$ C_x and (C_y, C_z) can always be found to yield two distinct s_d and s'_d each consisting of d columns. □

Let $S(d, n)$ be the sample space consisting of all samples with n items and d positives. Consider an $t \times n$ d-separable matrix. For each sample $s \in S(d, n)$, we can obtain t test outcomes which form a t-dimensional column vector, called a *test outcome vector*. Given a test outcome vector, how to decode it to recover the sample s, i.e., how to identify all positives?

First, we note that all items contained in a negative test can be identified to be negative immediately. The rest is to find d items among the remaining items to hit all positive pools. (By a set hitting another set, we mean that the intersection of the two sets is nonempty.) This is exactly a hitting set problem as follows:

> HITTING SET PROBLEM: Given a set X and a collection $\mathcal{C}$ of subsets of X, find a minimum-cardinality subset Y of X such that $Y \cap C \neq \emptyset$ for all $C \in \mathcal{C}$.

Let M_D be a $t_D \times n_D$ matrix obtained from M by keeping only the set T_D of positive pools and the set N_D of n_D items not appearing in any negative pool, given D the set of positives. Then the decoding is associated with a hitting problem with $X = N_D$ and $\mathcal{C} = T_D$. We refer to the hitting set problem obtained from the above interpretation as the *reduced hitting set problem.* The next lemma shows that if n_D, then the d positive items yield the unique minimum solution.

Lemma 2.1.2 *Given a $t \times n$ d-separable matrix and a test outcome vector resulting from a sample in $S(d,n)$, there exists a unique minimum solution for the reduced hitting set problem. Moreover, its size is d, except for $n_D = d$, the size can be $d-1$.*

Proof. Since N_D contains all positive items, we have $n_D \geq d$.

If $n_D > d$, we show that any minimum solution has size at least d. For contradiction, suppose there exists a hitting set H of size $h < d$. Then putting other $d-h$ items from N_D into H would result in a hitting set of size d. Since $n_D > d$, we can find, in this way, two distinct hitting sets of size d. Note that the union of columns corresponding to any hitting set is the test outcome vector. Therefore, the two unions corresponding to two hitting sets of size d are equal, contradicting the definition of d-separability.

Moreover, all d positive items form a hitting set for positive pools. Therefore, the minimum hitting set has size exactly d. Furthermore, the hitting set of size d is unique since existence of two distinct hitting sets of size d yields the equality of two unions of d columns, contradicting the d-separability.

When $n_D = d$, the minimum hitting set K may have size k smaller than d, which would not result in any contadiction since N_D has the unique subset of size d. By Lemma 2.1.1, a d-separable matrix is also $(d-1)$-separable. If $k < d-1$, then $K \cup \{x\} \neq K \cup \{y\}$ for $x \neq y$, $x, y \in N_D - K$, contradicting the $(d-1)$-separability. Hence, $k = d-1$. The d-separability also assures the uniqueness of K. □

The *weight* of a binary vector is the number of 1s in the vector.

Theorem 2.1.3 *In a d-separable matrix, the union of any d columns not including a zero column has weight at least d.*

Proof. Consider a $t \times n$ d-separable matrix. Let s be a subset of d columns not including a zero column. s can also be seen as a set of d items. Consider the case that s is exactly the set of d positive items. Note that each column C in a unique solution V must appear in a row in which no other column in V appears, or $V \setminus \{C\}$ would also be a solution, contradicting the minimality of V. Let R_V denote the set of such $|V|$ rows. Then $|U(s)| \geq |R_V| = |V|$. By Lemma 2.1.2, either s is the unique minimum solution, or s is the set of all items not appearing in any negative pool.

Suppose s is the unique solution of the hitting set problem resulting from the decoding. Then $U(s)$ has weight at least d since, otherwise, a hitting set of size at most $d-1$ exists.

Suppose s is the set of items not appearing in any negative pool. Let s' be the unique minimum solution of the reduced hitting set problem and $|s'| = d-1$. Moreover, for each column C_j in s', there exists a positive pool R_{i_j} hit only by C_j but not other columns in s. (If such a positive pool does not exist, then C_j can be removed from the hitting set, which results in a hitting set with cardinality $d-1$, different from s'.) Suppose the column C_k in $s \setminus s'$ hits a pool R_{i_j}. Then, replacing C_j in s' by C_k results in another hitting set of size $d-1$, a contradiction. Therefore, C_k does not hit any R_{i_j} for $C_j \in s'$. Since C_k is not a zero column, C_k must hit a row other than R_{i_j}'s. It follows that $U(s)$ has weight at least d. □

Corollary 2.1.4 *In a d-separable matrix, the union of any d columns has weight at least $d-1$.*

Proof. If the d columns include the zero column, apply Lemma 2.1.3 to the other $d-1$ columns. □

The hitting set problem is NP-hard. Even the size of the minimum hitting set is known, no clever method has been found so far to compute the minimum solution. The only way we know is an exhaustive search, that is, checking all possible d-subsets of unidentified items in time $O(|X|^d)$. Is there an algorithm of polynomial time with respect to both $|X|$ and d? The answer is NO if NP≠P. In fact, if such a polynomial-time algorithm exists, then we may run the algorithm $|X|$ times on inputs with $d = 1, 2, ..., |X|$ to solve the hitting set problem in polynomial time. This implies NP=P [18].

We now give corresponding results for $\bar{d}$-separable matrices whose proofs are usually simpler.

Lemma 2.1.5 *If a matrix is $\bar{d}$-separable, then it is $\bar{k}$-separable for all $1 \leq k \leq d \leq n$.*

Proof. By definition of $\bar{d}$-separable. □

No $\bar{d}$-separable matrix can contain a zero column. The following lemma states a relationship between d-separable, $\bar{d}$-separable, and the zero column.

Lemma 2.1.6 *A binary matrix containing a zero column is d-separable if and only if all nonzero columns form a $\bar{d}$-separable matrix.*

Proof. Let M be a binary matrix containing a zero column C and M' a submatrix obtained from M by deleting C. Suppose M' is $\bar{d}$-separable. Consider two samples s and s' in $S(d,n)$. Let s^* and s'^* be obtained respectively from s and s' by deleting C. Then $s^* \neq s'^*$. By the $\bar{d}$-separability of M', $U(s) = U(s^*) \neq U(s'^*) = U(s')$.

Conversely, suppose M is d-separable. Consider two samples s and s' in $S(\bar{d}, n-1)$, both consisting of nonzero columns. If $|s| = |s'|$, then by Lemma 2.1.1, $U(s) \neq U(s')$. Otherwise, without loss of generality, we may assume $|s| < |s'| \leq d$. Let $G \subset s' \setminus s$ and $|G| = |s'| - |s| - 1$. Suppose to the contrary that $U(s) = U(s')$. Then $U(s \cup G) = U(s \cup G) = U(s)$. Let s^* be obtained from $s \cup G$ by adding the zero column. Then $|s^*| = |s'|$ and $U(s^*) = U(s')$, but $s^* \neq s'$, contracting the d-separability of M. □

To decode test outcomes from a $\bar{d}$-separable matrix, we may first identify all items contained by negative pools. Next, set X to be the set of items not appearing in any negative pool and set $\mathcal{C}$ to be the collection of all positive pools. Then, all positive items form a hitting set of size at most d.

Let $S(\bar{d}, n)$ be the sample space consisting of all samples with n items and at most d positives. The following lemma indicates that all positive items actually form a unique hitting set of size at most d.

Lemma 2.1.7 *For any $t \times n$ $\bar{d}$-separable matrix and any test outcome vector resulting from a sample in $S(\bar{d}, n)$, there exists exactly one hitting set of size at most d for the reduced hitting set problem.*

Proof. Similar to the proof of Lemma 2.1.2. □

Corollary 2.1.8 *A $\bar{d}$-separable matrix does not contain a zero column.*

Proof. Suppose a $\bar{d}$-separable matrix contains a zero column indexed by item j. Then the two sample points $s = \emptyset$ and $s' = \{j\}$ have the same outcome $U(s) = U(s') = \emptyset$, contradicting the $\bar{d}$-separability. □

Note that a $\bar{d}$-separable matrix is also d-separable. Then by Theorem 2.1.3 and Corollary 2.1.8, we have

Lemma 2.1.9 *In any $\bar{d}$-separable matrix, the union of any d columns has weight at least d.*

Since the reduced hitting set problem is NP-hard in the input N_D, it is important to establish n_D. Chen, Hwang and Li [3] observed

Lemma 2.1.10 *If M is $d(\bar{d})$-separable, so is M_D.*

Proof. A column C in N_D preserves all its 1-entries in M_D. Thus if $U(D)$ does not equal the union of any other set of d (up to d) columns in M, it certainly does not in M_D. □

Thus, any bound of n for a $d(\bar{d})$-separable matrix M can be used to bound n_d, the number of columns in M_D.

The following is a characterization of n_D.

Lemma 2.1.11 *Consider a d-separable matrix M. $n_D \leq d+k-1$ for all $D \in S(d,n)$ if and only if the union of any k columns cannot be contained by the union of other d columns.*

Proof. Note that an item does not appear in a negative pool if and only if its corresponding column is contained by the union of d positive columns. Therefore, the number of items not appearing in any negative pool is more than $d+k$ if and only if there are at least k non-positive items whose columns are contained by the d positive columns. □

2.2 d-Disjunct Matrices

Consider a $t \times n$ binary matrix M where R_i and C_j denote row i and column j. M is said to be *d-disjunct* if the union of any d columns does not contain any other column.

Lemma 2.2.1 *If a matrix is d-disjunct, then it is k-disjunct for all $1 \leq k \leq d \leq n$.*

Bassalygo (see [7]) proved

Lemma 2.2.2 *Deleting a column and all row intersecting it from a d-disjunct matrix yields a $(d-1)$-disjunct matrix.*

Proof. Suppose to the contrary that there exists a column C and a set s of $d-1$ other columns in the reduced matrix such that $C \subseteq U(s)$. Then $C \subseteq U(s \cup \{\text{the deleted column}\})$ in the original matrix, contradicting d-disjunctness. □

Theorem 2.2.3 *A binary matrix is d-disjunct if and only if for any sample $s \in S(\bar{d},n)$, the union of negative pools contains all negative items.*

Proof. Suppose $s \in S(k,n)$ for $1 \leq k \leq d$. By defintion of d-disjunct and Lemma 2.1.11, we have $n_s \leq k$. Therefore, $n_s = k$, i.e., all items not appearing in any negative pool are positive. □

Theorem 2.2.3 implies that a d-disjunct matrix solves the $(\bar{d},n)$ problem with a simple decoding, i.e., all items not in negative pools constitute the set of positive items. Hence,

Corollary 2.2.4 *d-disjunct implies $\bar{d}$-separable (hence d-separable).*

Saha and Sinha [24] showed that d-disjunctness requires at least one more test than d-separability.

Lemma 2.2.5 *Deleting any row R_i from a d-disjunct matrix M yields a d-separable matrix M_i.*

Proof. Let $s, s' \in S(\bar{d}, n)$. Let $U(s)$ denote the set of positive pools under the sample s. Then $U(s)$ and $U(s')$ must differ in at least 2 rows or one would contain the other. Hence, they are different even after the deletion of a row. □

We extend this result to $\bar{d}$-separable under an extra condition.

Lemma 2.2.6 *Suppose M is a d-disjunct matrix with no isolated column. Then deleting any row from M yields a $\bar{d}$-separable matrix.*

Proof. Let $s, s' \in S(\bar{d}, n)$. If none contains the other, the argument is the same as in Lemma 2.2.4. So, assume $s \subset s'$. Let $C' \in s' - s$. Then $C' \not\subseteq U(s)$ by d-disjunctness. If $|C' \setminus U(s)| = 1$, let C be a column intersecting the row in $C' \setminus U(s)$ (the existence of C is assured by the assumption of no isolated column). Then $C' \subseteq s \cup \{C\}$, contradicting d-disjunctness again since $|s| < d$ and hence $|s \cup \{C\}| \leq d$. Therefore, $|C' \setminus U(s)| \geq 2$. This means that after deleting a row, we still have $U(s) \neq U(s')$. □

Kautz and Singleton proved

Lemma 2.2.7 $\overline{(d+1)}$*-separable implies d-disjunct.*

Proof. Suppose that M is $\overline{(d+1)}$-separable but not d-disjunct, i.e., there exists a sample s of d columns such that $U(s)$ contains another column C_j not in s. Then

$$U(s) = C_j \cup U(s),$$

a contradiction to the assumption that M is $\overline{(d+1)}$-separable. □

Thus, a $\overline{(d+1)}$-separable matrix is both $(d+1)$-separable and d-disjunct. Chen and Hwang [2] proved the backward direction is also true.

Lemma 2.2.8 *A binary matrix is $\overline{(d+1)}$-separable if and only if it is $(d+1)$-separable and d-disjunct.*

Proof. It suffices to show only the backward direction. Consider two samples s and s' from $S(\overline{d+1}, n)$. If $|s| = |s'| = d+1$, then $U(s) \neq U(s')$ by $(d+1)$-separability. Else, assume $|s| \leq |s'|$. Then $|s| \leq d$ and s' must contain a column C_j not in s. By d-disjunctness, $U(s)$ does not contain C_j and hence $U(s) \neq U(s')$. □

From Corollary 2.2.4 and Lemma 2.2.8 any property held by a $\bar{d}$-separable matrix also holds for a d-disjunct matrix, and any property held for a d-disjunct matrix also holds for a $\overline{d+1}$-separable matrix.

Theorem 2.2.9 *Let M be a $2d$-separable matrix. Then we can obtain a d-disjunct matrix by adding at most one row to M.*

Proof. If M is d-disjunct, then we are done. Suppose M is not. Then there exists a column C and a set s of d other columns such that C is contained by the union $U(s)$ of the d columns in s. Add a row R which has 1-entry at C and 0-entries at all columns in s to break up the containment $C \subset U(s)$ in the new matrix. However, there may exist many pairs (C, s) such that $C \subset U(s)$. If we use the same row to break up the containment, then we must show that there is no conflict in the definition of R, that is, we must show that for two such pairs (C, s) and (C', s'), we would not have $C \in s'$.

For contradiction, suppose $C \in s'$. Set $S_0 = s \cup s' \cup \{C, C'\}$. Then $|S_0| \leq 2d+1$. Set $S_1 = S_0 \setminus \{C\}$ and $S_2 = S_0 \setminus \{C'\}$. Then $|S_1| = |S_2| \leq 2d$. However, $U(S_1) = U(S_0) = U(S_2)$, contradicting $2d$-separability. □

Corollary 2.2.10 *Let M be a $2d$-separable matrix. Then we can obtain a $\overline{(d+1)}$-separable matrix by adding at most one row to M. In particular,*

$$t_s(\bar{2}, n) - t_s(2, n) \leq 1,$$

where $t_s(\bar{2}, n)$ ($t_s(2, n)$) is the minimum number of rows for a 2-separable ($\bar{2}$-separable) matrix to have n columns.

Proof. By Theorem 2.2.9 (set $d = 1$) and Lemma 2.2.8. □

2.3 The Minimum Number of Pools for Given d and n

For a given set P of parameters, let $t(P)$ denote the minimum number of rows required for a disjunct matrix satisfying P and $n(P)$ the maximum number of columns required. Similarly, $t_s(P)$ and $n_s(P)$ will denote the same for a separable matrix. For example $t(d, n)$ denotes the minimum number of rows required for a d-disjunct matrix with n columns, while $n_s(\bar{d}, t)$ denotes the maximum number of columns for a $\bar{d}$-separable matrix with t rows. In this section we derive some basic results about these functions.

Lemma 2.3.1 *For a d-separable matrix M and $s \in S(d, n)$, $|U(s)| \geq d-1$. If s contains no 0-column, then $|U(s)| \geq d$.*

Proof. s contains at least $d-1$ nonzero columns. By Corollary 2.1.3, $|U(s)| \geq d-1$. The second part is an alternative statement of Corollary 2.1.3. □

Corollary 2.3.2 *For a d-separable $t \times n$ matrix*

$$\binom{n}{d} \leq \sum_{i=d-1}^{t} \binom{t}{i}.$$

For example, $n_s(3,5) \leq 6$ since

$$\binom{6}{3} \leq \sum_{i=2}^{5} \binom{5}{i} = 26 < \binom{7}{3}.$$

Corollary 2.3.3 $t_s(d,n) \geq d \log n(1+o(1))$ *for* $n >> d$.

Proof. By using the approximation

$$\binom{t}{d} \sim (\frac{et}{d})^d$$

and the fact

$$\sum_{i=d}^{t} \binom{t}{i} \leq 2^t.$$

□

Recall that a column is called isolated if there exists a row incident to it but not to any other column. Since the removal of an isolated column and its incident row does not affect the testing of other columns, we often assume the matrix in study has no isolated column to facilitate the analysis.

Lemma 2.3.4 *A nonisolated column in a d-disjunct matrix has at least* $d-k+1$ *1-entries not covered by any other* k *columns,* $0 \leq k \leq d$.

Proof. Suppose to the contrary that there exist a nonisolated column C and a set K of k columns which covers C except for at most $d-k$ 1-entries. Since each uncovered 1-entry of C must appear in another column, at most $d-k$ columns additional to K cover C, violating the assumption of d-disjunctness. □

The special case of $k=0$ was first proved by Dyachkov and Rykov [6], which is stated as a corollary.

Corollary 2.3.5 *A nonisolated column in a d-disjunct matrix has weight at least* $d+1$.

For a given $t \times n$ binary matrix, let $c(w)$ denote its number of columns with weight w. Then

$$n = \sum_{w=0}^{t} c(w).$$

Lemma 2.3.6

$$\sum_{w=0}^{d} c(w) \leq t.$$

Proof. By Corollary 2.3.5, each column with weight $\leq d$ is isolated. Lemma 2.3.6 follows since each isolated column consumes a row. □

Bassalygo (see [7]) proved the first nontrivial lower bound of $t(d,n)$.

Theorem 2.3.7 $t(d,n) \geq \min\left\{\binom{d+2}{2}\ ,\ n\right\}$ *for a d-disjunct matrix.*

Proof. Trivially true for $n = 1$. The general case is proved by induction on n.

Let M denote a d-disjunct matrix achieving $t(d,n)$. Suppose M has a column of weight $w \geq d+1$. Then by Lemma 2.2.2,

$$\begin{aligned} t(d,n) &\geq d+1+t(d-1,n-1) \\ &\geq d+1+\min\left\{\binom{d+1}{2}, n-1\right\} \\ &\geq \min\left\{\binom{d+2}{2}, n\right\}. \end{aligned}$$

Suppose M does not have such a column. Then by Lemma 2.3.6,

$$\begin{aligned} t(d,n) &\geq \sum_{w=0}^{d} n(w) = n \\ &\geq \min\left\{\binom{d+2}{2}, n\right\}. \end{aligned}$$

□

Corollary 2.3.8 *For* $\binom{d+2}{2} \geq n$, $t(d,n) \ = \ n$.

Erdös, Frankl and Füredi [9, 10] conjectured

$$t(d,n) = n \text{ if } n \leq (d+1)^2,$$

and proved for $d \leq 3$. Huang and Hwang [13] proved for $d = 4$, while Chen and Hwang [2] for $d = 5$.

Kautz and Singleton [17] introduced some secondary parameters. Let w_j denote the weight of C_j and let λ_{ij} denote the dot product of C_i and C_j, i.e., the number of rows that both columns have a 1-entry (we also say C_i and C_j intersect λ_{ij} times). Define

$$\underline{w} = \min_j w_j$$

and

$$\bar{\lambda} = \max_{i,j} \lambda_{ij}.$$

They gave a fundamental result which is the basis of many constructions.

Lemma 2.3.9 *A binary matrix is d-disjunct where*

$$d = \lfloor \frac{\underline{w} - 1}{\bar{\lambda}} \rfloor.$$

Proof. A column C has at most $d\bar{\lambda}$ intersections with the union of any d other columns. Note that

$$d = \lfloor \frac{\underline{w} - 1}{\bar{\lambda}} \rfloor$$

implies $\underline{w} \geq d\bar{\lambda} + 1$. Hence C is not covered by the union. □

Let $\bar{\lambda}_k(\underline{\lambda}_k)$ denote the maximum (minimum) number of intersections of a set of k columns. Then Lemma 2.3.9 can be generalized to

Lemma 2.3.10 *A binary matrix is d-disjunct if*

$$\binom{d}{0}\underline{\lambda}_1 - \binom{d}{1}\bar{\lambda}_2 + \binom{d}{2}\underline{\lambda}_3 - \binom{d}{3}\bar{\lambda}_4 + \cdots - \binom{d}{2k-1}\bar{\lambda}_{2k} > 0 \text{ for some } 1 \leq k \leq \lfloor d/2 \rfloor.$$

Proof. The left-hand side is a truncated inclusion-exclusion formula to compute the number of 1-entries in a column not covered by some other d columns. The real number is at least as large as the truncated number since the first truncated term is positive and also since lower bounds of λ_k are used for all positive terms and an upper bounds for all negative terms. □

2.4 Combinatorial Bounds for d-Disjunct Matrices with Constant Weight

For matrices with constant column weight w, we denote n by $n(w)$ and $n(d,t)$ by $n(d,t,w)$. In this section we give some bounds of the $n(w)$ function which provide the tool to bound $t(d,n)$. For matrices with different weights, then

$$n(d,t) \leq \sum_{w=0}^{t} n(d,t,w).$$

In particular, if the weights are bounded by $\bar{w}$, then

$$n(d,t) \leq \sum_{w=0}^{\bar{w}} n(d,t,w).$$

A very important notion in studying constant-weight matrix is "privateness". For a given matrix M, a subset of $\{1,...,t\}$ is *private* if it belongs to a unique column.

Johnson [16] observed

Lemma 2.4.1 *$n(d,t,w) \leq \binom{t}{\bar{\lambda}+1}/\binom{w}{\bar{\lambda}+1}$ where $\bar{\lambda}$ is the maximum size of intersection of two columns.*

Proof. By the definition of $\bar{\lambda}$, any $(\bar{\lambda}+1)$-subset of a column is private. Since there are $\binom{t}{\lambda+1}$ $(\bar{\lambda}+1)$-subsets in total, while a column has $\binom{w}{\bar{\lambda}+1}$ $(\bar{\lambda}+1)$-subsets, the ratio bounds the number of columns. □

Frankl [11] proved

Lemma 2.4.2 *A column in a d-disjunct matrix with constant column weight w has at most*

$$\binom{w-1}{v}$$

nonprivate v-subsets for any $v \geq \lceil w/d \rceil$.

Proof. It suffices to prove for $v = \lceil w/d \rceil$. Let C be a column and let $F(C)$ denote the family of nonprivate v-subsets in C. Consider a cyclic ordering $\pi = (\pi_1, \pi_2, ..., \pi_w)$ of the labels of the w rows incident to C. Define $S_i = (\pi_{i-v+1}, \pi_{i-v+2}, ..., \pi_i)$ for $1 \leq i \leq w$. Note that there cannot exist d S_i in $F(C)$ such that their union is C, for there would exist at most d columns containing all these S_i (since each S_i is not private), and hence the union of the d columns cover C, contradicting the assumption of d-disjunctness. We prove that under this condition, at most $w - v$ S_i can lie in $F(C)$.

Without loss of generality, assume $S_w \in F_v$. Partition the S_i into v classes where class $C_j = \{S_i \mid i \equiv j \pmod{v}\}$. Define $d' = \lceil w/v \rceil$. Suppose

$$w = d'v - q, \quad 0 \leq q < v - 1.$$

Then

$$|C_j| = \begin{cases} d' & \text{for } 1 \leq j \leq v-q \\ d'-1 & \text{for } v-q+1 \leq j \leq v. \end{cases}$$

Note that for those C_j with $|C_j| = d'$, their union is clearly $\{1, ..., w\}$. The union of C_j is equal to $\{\pi_1, \pi_2, ..., \pi_{w-v+j-q}\}$ for $q+1 \leq j \leq v$. Hence its union with $S_w = \{\pi_{w-v+1}, \pi_{w-v+2}, ..., \pi_w\}$ is also $\{1, ..., w\}$. Therefore, every class must exclude at least one S_i from $F(C)$, i.e., at most $w - v$ S_i can be in $F(C)$.

There are $(w-1)!$ cyclic ordering, each having at most $w-v$ consecutive v-subsets in $F(C)$. On the other hand, each v-subset in $F(C)$ can be in $v!(w-v)!$ cyclic ordering. So counting all the nonprivate v-subsets in all possible cyclic ordering in two ways, we obtain

$$(w-1)!(w-v) \geq |F(C)|v!(w-v)!,$$

or

$$|F(C)| \leq \frac{(w-1)!(w-v)}{v!(w-v)!} = \binom{w-1}{v}.$$

□

Erdös, Frankl and Füredi [9, 10] proved

Theorem 2.4.3 $n(d,t,w) \leq \binom{t}{v}/\binom{w-1}{v-1}$ *where* $v = \lceil w/d \rceil$.

Proof. By Lemma 2.4.2, each column of weight w has at least

$$\binom{w}{v} - \binom{w-1}{v} = \binom{w-1}{v-1}$$

private v-subsets. By the definition of privateness, the private v-subsets of different columns are disjoint. Since there are only $\binom{t}{v}$ distinct v-subsets, there are at most

$$\binom{t}{v}/\binom{w-1}{v-1}$$

columns of weight w. □

Corollary 2.4.4 $n(d,t,w) \leq d\binom{t}{v}/\binom{w}{v}$.

Note that Theorem 2.4.3 holds for all $v \geq \lceil w/d \rceil$, but the tightest bound is given by $v = w/d$ for $t \geq 2w$.

D'yachkov and Rykov [6, 7] independently proved a slightly weaker version.

Corollary 2.4.5 $n(d,t,w) \leq d^2\binom{t}{\lceil w/d \rceil}/\binom{d\lceil w/d \rceil}{\lceil w/d \rceil}$.

Erdös, Frankl and Füredi also proved a lower bound.

Theorem 2.4.6 $n(w) \geq \binom{t}{v}/\binom{w}{v}^2$.

Proof. Let $\binom{[t]}{w}$ denote the set of binary t-vector with constant weight w, and $\mathcal{P}$ a maximal set of $\binom{[t]}{w}$ such that two vectors intersect at most $v-1$. ($\mathcal{P}$ is called a

(v, w, t)-packing in Section 3.1.) For each member V of $\binom{[t]}{w}$, there exists a U in $\mathcal{P}$ such that $|U \cap V| \geq v$ (if $V \in \mathcal{P}$, set $U = V$). Then

$$\sum_{U \in \mathcal{P}} |\{V \mid U \cap V| \geq v\}| \geq |\binom{[t]}{w}| = \binom{t}{w}.$$

On the other hand, for a given U with its fixed w 1-entries, any V containing v of these w 1-entries and $w - v$ other 1-entries is in the set. Hence

$$\sum_{U \in N} |\{V \mid U \cap V| \geq v\}| \leq |\mathcal{P}| \binom{w}{v} \binom{t-v}{w-v}.$$

Note that a V can be counted many times on the right-hand side, thus the inequality. From these two inequalities, we conclude

$$|\mathcal{P}| \geq \binom{t}{w} / \binom{w}{v} \binom{t-v}{w-v} = \binom{t}{v} / \binom{w}{v}^2.$$

□

Hwang and Sös [15] strengthened the bound slightly.

Corollary 2.4.7 $n(w) \geq \binom{t}{w} / \sum_{i=v}^{w} \binom{w}{i} \binom{t-w}{w-i}$.

Proof. Let $\binom{[t]}{w}$ denote the set of t-vectors with weight w, and let F_v be a maximal subset of $\binom{[t]}{w}$ consisting of columns not intersecting more than $v - 1$. A column $C \in F_v$ has at most

$$\sum_{i=v}^{w} \binom{w}{i} \binom{t-w}{w-i}$$

columns in $\binom{T}{w}$ intersecting at least v with C, hence being excluded from F_v. Therefore

$$n(w) \geq |F_v| \geq \binom{t}{w} / \sum_{i=v}^{w} \binom{w}{i} \binom{t-w}{w-i}.$$

□

2.5 Asymptotic Lower and Upper Bounds

Due to Corollary 2.2.3 and Theorem 2.2.9, asymptotic results can be interchanged among the three classes of matrices, d-separable, $\bar{d}$-separable and d-disjunct with a possible adjustment of coefficients. Let $B(d)$ denote a bound. In particular,

$$\begin{aligned}
t(d,n) \leq B(d) &\Rightarrow t_s(d,n) \leq t_s(\bar{d},n) \leq B(d), \\
t_s(d,n) \leq B(d) &\Rightarrow t(d,n) \leq B(2d) \text{ and} \\
&\quad\;\; t_s(\bar{d},n) \leq \max\{t_s(d,n), t(d-1,n)\} \leq B(2d-2), \\
t_s(d,n) \geq B(d) &\Rightarrow t(d,n) \geq t_s(\bar{d},n) \geq B(d), \\
t(d,n) \geq B(d) &\Rightarrow t_s(\bar{d},n) \geq t_s(d,n) \geq B(d/2).
\end{aligned}$$

In this section, we will present our results only for one type of separating matrices (mostly the disjunct type) and rely on the above relations to extend to the other types.

Let $h(x)$ denote the entropy function, i.e.,

$$h(x) = -x \log x - (1-x)\log(1-x).$$

Then

$$h(x) \to -x \log x \text{ as } x \to 0.$$

By using the Stirling formula, we have

$$\binom{t}{\lambda t} \sim 2^{th(\lambda)}.$$

This approximation will be used in approximating $\binom{t}{v}/\binom{w}{v}$.

Lemma 2.5.1 *Define* $\alpha = dw/t$. *Assume* $\alpha << d^2$. *Then*

$$\binom{t}{w/d} \Big/ \binom{w}{w/d} \sim 2^{(\alpha/d^2)t \log d}.$$

Proof.

$$\begin{aligned}\frac{\binom{t}{v}}{\binom{w}{v}} &\sim \frac{\binom{t}{(\alpha/d^2)t}}{\binom{w}{(1/d)w}} \\ &\sim \frac{2^{th(\alpha/d^2)}}{2^{wh(1/d)}} = 2^{t[h(\alpha/d^2)-(\alpha/d)h(1/d)]} \\ &\sim 2^{t[(\alpha/d^2)\log(d^2/\alpha)-(\alpha/d)(1/d)\log d]} \\ &\sim 2^{(\alpha/d^2)t\log d}.\end{aligned}$$

□

Corollary 2.5.2

$$t(d,n,w) \geq \frac{d^2 \log n}{\alpha \log d}(1+o(1)).$$

Proof. By Corollary 2.4.4 and Lemma 2.5.1,

$$n(d,t,w) \leq \frac{d\binom{t}{v}}{\binom{w}{v}} \leq d2^{(\alpha/d^2)t\log d}(1+o(1)).$$

Corollary 2.5.2 follows by taking log on both sides. □

D'yachkov and Rykov [6] used a recursive method to compute α.

Define $K_d = d^2/\alpha \log d$ and $f_d(\beta) = h(\beta/d) - \beta h(1/d)$. D'yachkov and Rykov gave a formula to determine the proper weight w and to compute

$$K_1 = 1,$$
$$K_2 = \max_{0<\beta<1}(1/f_2(\beta)),$$
$$K_d = 1/f_d((K_d - K_{d-1})/K_d) \text{ for } d \geq 3.$$

In particular, they found $K_2 = 3.106$, $K_3 = 5.018$, $K_4 = 7.120$, $K_5 = 9.416$, $K_6 = 12.048$, and in general:

Theorem 2.5.3

$$K_d \geq d^2/2\log[e(d+1)/2] \textit{ for } d \geq 2.$$

In particular,

$$K_d \geq \frac{d^2}{2\log d}(1+o(1)) \textit{ for } d \textit{ large.}$$

For d-separable $K_d \geq d$ for $d \geq 11$ (private communication from D'yachkov and Rykov).

Corollary 2.5.4

$$t(d,n) \geq \frac{d^2 \log n}{2\log d}(1+o(1)).$$

This is the best lower bound so far. Note that this bound is derived from a constant-weight matrix, though the optimal weight is determined only recursively, and hence numerically.

Nguyen and Zeisel [21] gave an argument that $\alpha < 2$ in Corollary 2.5.2 by using the Johnson bound [16]

$$t(w-\bar{\lambda}) \geq n(w^2 - t\bar{\lambda}).$$

They noted

$$\begin{aligned} w^2 - t\bar{\lambda} &\geq (\frac{\alpha t}{d})^2 - \frac{tw}{d} \geq (\frac{\alpha t}{d})^2 - \frac{\alpha t^2}{d^2} \\ &= \frac{\alpha t^2}{\lambda^2}(\alpha - 1) > 0 \text{ if } \alpha > 1. \end{aligned}$$

So, under the assumption $\alpha > 1$, the Johnson bound can be written as

$$n \leq \frac{t(w-\bar{\lambda})}{w^2 - t\bar{\lambda}},$$

which is increasing in $\bar{\lambda}$. Therefore, setting $\bar{\lambda}$ to its maximum value w/d and $w = \alpha t/d$, we obtain

$$n \leq \frac{t(\alpha t/d - \alpha t/d^2)}{(\alpha t^2/d^2)(\alpha - 1)} = \frac{d-1}{\alpha - 1}.$$

Since $n \geq d - 1$, necessarily $\alpha < 2$.

In fact, they argued that $\alpha < 1$ for $n \to \infty$. This is because given an $\alpha > 1$, then there exists an n large enough so that $n \leq (d-1)/(\alpha - 1)$ does not hold. Since this inequality is derived from the assumption $\alpha > 1$, the assumption must be refuted for $n \to \infty$.

Ruszinsko [23] gave a combinatorial argument that $\alpha \leq 8$ in Corollary 2.5.2. Füredi [12] improved to $\alpha \leq 4$.

Theorem 2.5.5 $t(d, n) \geq (d^2 \log n / 4 \log d)(1 + o(1))$.

Proof. Let M be a $t \times n$ d-disjunct matrix and F its set of columns. Set $v = \lceil (t-d)/\binom{d}{2} \rceil$. Let F_t denote the set of columns each containing a private v-subset, F_0 the set of columns with fewer than v 1-entries, and $F' = F - F_t - F_0$. Let P denote the set of private v-subset contained in F_t, and let Q denote the set of v-subsets each containing a member of F_0. Then P and Q are disjoint since $p = q$, $p \in P$ and $q \in Q$ would imply a column in F_t contains a column in F_0, contradicting the assumption of d-disjunctness. Now, $|F_t| \leq |P|$ since two columns cannot share a private set, and $|F_v| \leq |Q|$ due to a lemma of Sperner [25]. It follows

$$|F_t| + |F_0| \leq |P| + |Q| \leq \binom{t}{v}.$$

Let $C_0 \in F'$ and $C_1, ..., C_k \in F$, $1 \leq k \leq d$. Then

$$|C_0 \setminus \cup_{j=1}^{k} C_j| > v(d-k)$$

for if not, then all elements in $C_0 \setminus \cup_{j=1}^{k} C_j$ can be assigned to at most $d - k$ v-subsets (not necessarily disjoint) each contained in a column of F, contradicting the assumption of d-disjunctness.

Suppose $|F'| > d$. Then we can find $d+1$ columns $C'_0, C'_1, ..., C'_d$ in F' such that

$$\begin{aligned} |\cup_{j=0}^{d} C'_j| &= |C'_0| + |C'_1 \setminus C'_0| + |C'_2 \setminus (C'_0 \cup C'_1)| + \cdots + |C'_d \setminus (C'_0 \cup \cdots C'_{d-1})| \\ &> d + 1 + v(d-1) + v(d-2) + \cdots + v \\ &= d + 1 + v\binom{d}{2} > t, \end{aligned}$$

an absurdity since all columns are subsets of $\{1, ..., t\}$. Therefore $|F'| \leq d$. Thus

$$|F| = |F_t| + |F_0| + |F'| \leq \binom{t}{\lceil \frac{t-d}{\binom{d}{2}} \rceil} + d.$$

Theorem 2.5.5 follows immediately by using the approximation

$$\binom{t}{\lambda t} \sim 2^{th(\lambda)},$$

and

$$\lambda \sim 2/d^2,$$

i.e.,

$$n \leq 2^{t(2/d^2)\log(d^2/2)}(1+o(1)).$$

Hence,

$$\log n \leq \frac{2t}{d^2}\log d^2(1+o(1))$$

that is

$$t \geq [d^2 \log n / 4 \log d](1+o(1)).$$

□

D'yachkov, Rykov and Rashad [8] gave an upper bound of $t(d,n)$ using the random coding method which is too complicated to be explained here in detail. So, we just cite the result.

Theorem 2.5.6 *For n large,*

$$t(d,n) \leq (\log e) d^2 \log n(1+o(1)).$$

This bound is currently the best. D'yachkov and Rykov also indicated in a private communication that

$$t_s(d,n) \leq \frac{1}{2}(\log e) d^2 \log n(1+o(1)).$$

For $d=2$, Erdös, Frankl and Füredi gave an improvement over Theorem 2.4.3.

Theorem 2.5.7 *For $d=2$*

$$n(2,t,2v-1) \leq \binom{t}{v} \Big/ \binom{2v-1}{v}$$

$$n(2,t,2v) \leq \binom{t-1}{v} \Big/ \binom{2v-1}{v}.$$

Proof. Consider a column C of weight $2v$. A partition of C into two disjoint parts must contain a private part, or C would be covered by the union of two other columns. In particular, a partition of C into two v-parts implies the existence of a private v-part. Since there are $\frac{1}{2}\binom{2v}{v}$ such partitions of C, each column has at least that many private v-parts. Further, the private v-parts over all columns are distinct (by definition of privateness). Hence

$$n(2,t,2v-1) \leq \binom{t}{v} / (\frac{1}{2}\binom{2v}{v}) = \binom{t}{v} / \binom{2v-1}{v}.$$

We can do slightly better by replacing the numerator to the number of all private v-parts. It can be shown that at most $\binom{t-1}{v}$ v-parts are private which is obtained by having every $2v$-subset contains a fixed element x and specifying a v-part is private if and only if it doesn't contain x.

The bound of the case $w = 2v - 1$ is obtained from the above construction by deleting x. □

Next we investigate upper bounds of $t(d, n, w)$.

Using an analysis similar to the one in Lemma 2.5.1 on $\binom{t}{v} / \binom{w}{v}^2$ does not produce anything useful since the right-hand side becomes

$$2^{t[(\alpha/d^2)\log(d^2/\alpha) - 2(\alpha/d)(1/d)\log d]} \sim 2^0.$$

On the other hand, the improved bound of Lemma 2.4.7 does produce something useful by setting w and v properly.

Theorem 2.5.8

$$t(d,n) \leq 16(\log_3 2) d^2 \log n (1 + o(1)).$$

Proof. Set $b_i = \binom{w}{i}\binom{t-w}{w-i}$, $r = t/16d^2$ and $w = t/4d$. The for $3r < i \leq w$,

$$\frac{b_{i+1}}{b_i} = \frac{(w-i)^2}{(i+1)(t-2w+i+1)} \leq \frac{(4dr-3r)^2}{3r(t-8dr+3r)} \leq \frac{(4d-3)^2}{3(16d^2-8d+3)} < \frac{1}{3}.$$

Set $v = 4r$. Then

$$\begin{aligned} \sum_{i=v}^{w} b_i &< \sum_{i=4r}^{w} (\frac{1}{3})^{i-3r} b_{3r} \\ &< b_{3r} (\frac{1}{3})^{-3r} (\frac{1}{3})^{4r} / (1 - \frac{1}{3}) \\ &= b_{3r} (\frac{1}{3})^{r-1} / 2 \\ &< \binom{t}{w} (\frac{1}{3})^{r-1} / 2. \end{aligned}$$

By Lemma 2.4.7,

$$n(4dr) > \frac{\binom{t}{4dr}}{\sum_{4r<i<4dr} b_i} > 2 \cdot 3^{r-1} \geq 2 \cdot 3^{t/16d^2-1}.$$

Theorem 2.5.8 follows immediately. □

For some particular parameters, Erdös, Frankl and Furedi [10] showed that the asymptotic upper and lower bounds can match.

Dyachkov, Rykov and Rashad [8] obtained an asymptotic upper bound of $t(d,n)$.

Theorem 2.5.9 *For d constant and $n \to \infty$,*

$$t(d,n) \leq \frac{d}{A_d}(1+o(1))\log n \ ,$$

where

$$A_d = \max_{0\leq q\leq 1} \max_{0\leq Q\leq 1} \left\{ -(1-Q)\log(1-q^d) + d\left[Q\log\frac{q}{Q} + (1-Q)\log\frac{1-q}{1-Q} \right] \right\}.$$

They also showed that

$$A_d \to \frac{1}{d\log e}(1+o(1)) \text{ as } d \to \infty \ .$$

Let $\mathcal{P}$ be a (v,w,t)-packing. Rödl [22] proved

Lemma 2.5.10

$$\max\{|\mathcal{P}| \mid \mathcal{P} \textit{ is a } (v,w,t)\textit{-packing}\} = (1-o(1))\frac{\binom{t}{v}}{\binom{w}{v}}\}$$

for fixed w and v when $t \to \infty$.

Theorem 2.5.11 *Let $w = d(v-1)+1+q$ where $0 \leq q < d$. Then for n large,*

$$(1-o(1))\frac{\binom{t-q}{v}}{\binom{w-q}{v}} \leq n(d,t,w) \leq \frac{\binom{t-q}{v}}{\binom{w-q}{v}}$$

holds in the following cases:

(i) $q = 0, 1$,

(ii) $q < d/2v^2$,

(iii) $v = 2$ *and* $q < \lceil 2d/3 \rceil$.

Proof. Lower bound. We first prove it for $q = 0$. Consider $\mathcal{P}$ as defined in Theorem 2.5.5 with parameter (v, w, t). Rödl proved (see Theorem 3.1 in [22])

$$\max |\mathcal{P}| = (1 + o(1)) \frac{\binom{t}{v}}{\binom{w}{v}}.$$

Set $w = d(v - 1) + 1$. By Lemma 2.3.9, $\mathcal{P}$ is d-disjunct. Hence the lower bound is proved.

For $1 \leq q \leq d$, let $\mathcal{P}'$ be obtained by adding q new elements to t. These q elements also appear in every w-subset of $\mathcal{P}$. Thus $\mathcal{P}'$ has parameters (v, w', t') with $w' = w + q$ and $t' = t + q$. Further, $\mathcal{P}'$ preserves d-disjunctness and the number of w'-subsets equals that of w-subsets in $\mathcal{P}$. Hence the lower bound holds.

Upper bound. As the arguments are complicated, we refer the reader to [22]. □

2.6 (d, r)-Disjunct Matrices

D'yachkov and Rykov [7] extended the notion of d-disjunctness to (d, r)-disjunctness, which they called a *superimposed (d, n, r)-code of length t*, if for $d + r$ arbitrary columns, the union of any r columns is not contained in the union of the other d columns, i.e., there exists a row in which one of the r columns appears but none of the d columns does. Note that $(d, 1)$-disjunct is just d-disjunct. They proved

Lemma 2.6.1 *For $d' > d$ and $r' > r$, (d', r)-disjunct implies (d, r)-disjunct which implies (d, r')-disjunct. Hence*

$$t(d', r, n) \geq t(d, r, n) \geq t(d, r', n).$$

Lemma 2.6.2 *In a (d, r)-disjunct matrix, at most $\binom{d+r}{d}$ d-subsets of columns can have a union equal to the union of a given d-subset. Hence*

$$t(d, r, n) \geq \lceil \log \binom{n}{d} - \log \binom{d+r}{d} \rceil.$$

Lemma 2.6.3 $t(d, r, n) \geq t(\lfloor d/(r+1) \rfloor, 0, \lfloor n/(r+1) \rfloor)$ *for* $d > r \geq 1$.

Proof. Let M be a $t \times n$ (d, r)-disjunct matrix. Fixed n' $(= \lfloor n/(r+1) \rfloor)$ pairwise disjoint subsets $S_1, ..., S_{n'}$ of columns of M where each subset S_i contains at most $r+1$ columns. Let M' be $t \times n'$ binary matrix where each column corresponds to the union of a fixed S_i. Then in M', no column can be contained in the union of $\lfloor d/(r+1) \rfloor$ other columns since that would imply the union of a set of at most $r + 1$ columns is contained by a union of at most d columns, contradicting the (d, r)-disjunctness of M. □

De Bonis, Gasieniec and Vaccaro [4] obtained a stronger lower bound.

Theorem 2.6.4 *For* $n > (r+1)^2/(4d)$,

$$t(d,r,n) > \begin{cases} d\log(n/e(d+r)) & \text{for } 1 \leq d < 2(r+1), \\ \frac{(r+1)\lfloor d/(2r+2)\rfloor^2}{\log(ed^2/4(r+1))} \cdot \log \frac{4(n-2r-d/2)}{ed^2} & \text{for } d \geq 2(r+1). \end{cases}$$

Proof. For $d < 2(r+1)$, using

$$2^{y\log(x/y)} \leq \binom{x}{y} \leq 2^{y\log(ex/y)},$$

we obtain from Lemma 2.6.2

$$\begin{aligned} t(d,r,n) &\geq d\log(n/d) - d\log(e(d+r)/d) \\ &\geq d\log(n/e(d+r)). \end{aligned}$$

For $d \geq 2(r+1) \geq 4$, assume d is a multiple of $2(r+1)$ for easy presentation. Let M be a $t \times n$ (d,r)-disjunct matrix, and F its set of columns. We construct a $t' \times n'$ $(d/2,r)$-disjunct matrix M' with F' as its set of columns with maximum column weight $\lfloor 2t/d \rfloor$.

F' is obtained from F by deleting a column C with weight $> \lfloor 2t/d \rfloor$, and replacing each other column C' by $C' \setminus C$. Note that after ℓ ($\leq d/2$) such C-steps, all columns have weights $\leq \lfloor 2t/d \rfloor$. Further, at most r columns C' outside of C can disappear, i.e., $C' \setminus \cup_{j=1}^{\ell} C_j = \emptyset$ since otherwise the union of $r+1$ columns is contained in $\cup_{j=1}^{\ell} C_j$, contradicting the (d,r)-disjunctness of M. Therefore, $n' = |F'| \geq n - (d/2) - r$. Further, M' is $(d/2,r)$-disjunct since if there exists a set S' of $r+1$ columns whose union is contained in a set T' of $d/2$ columns in M', then $U(S) \subseteq (U(T)) \cup (\cup_{j=1}^{\ell} C_j)$ in M, where S and T are the corresponding columns of S' and T' in M and $\{C_j\}$ are the columns defining the ℓ steps, contradicting the d-disjunctness of M.

For each $(r+1)$-subset S' of F', there exists a subset T' of row indices with $|T'| \leq \lceil 4t(r+1)/d^2 \rceil$ such that $T' \subseteq C'$ for some $C' \in S'$, but $T' \not\subseteq C'$ for all $C' \notin S'$. Suppose not. Then we can partition each column in S' into at most $d/2(r+1)$ subsets of size at most $\lceil 4t(r+1)/d^2 \rceil$ such that each such subset is contained in a column in $F' \setminus S'$. Consequently, $U(S')$ is contained in a set of at most $(r+1)(d/2(r+1)) = d/2$ columns in $F' \setminus S'$, contradicting the $(d/2,r)$-disjunctness of M'.

Partition F', except at most r of them, into s $(r+1)$-sets where $s = \lfloor |F'|/(r+1) \rfloor$. Let T_i' denote the set T' defined in the above paragraph for the ith such $(r+1)$-set. Then $\{T_1', ..., T_s'\}$ is a Sperner family with $|T_i'| \leq \lceil 4t(r+1)/d^2 \rceil$ for all i. Since $t' \leq t$ and $\lceil 4t(r+1)/d^2 \rceil \leq \lceil 2t/d \rceil \leq \lceil t/2 \rceil$,

$$s \leq \binom{t}{\lceil 4t(r+1)/d^2 \rceil}$$

and

$$n \leq n' + r + d/2 \leq (r+1)\binom{t}{\lceil 4t(r+1)/d^2 \rceil} + 2r + d/2.$$

Solving for t, we obtain

$$t(d,r,n) > \frac{d^2}{4(r+1)\log(ed^2/4(r+1))} \log \frac{4(n-2r-d/2)}{ed^2}.$$

After taking into consideration that $d/2(r+1)$ may not be an integer, we obtain the formula in Theorem 2.6.4. □

Next, we give an upper bound of $t(d,r,n)$. Actually, we will introduce an upper bound of a stronger combinatorial object proposed by De Bonis, Gasieniec and Vaccaro [4]. A (k,m,n)*-selector* is a binary matrix with n columns such that for any k-set K of columns, at least m columns are isolated in K.

Lemma 2.6.5 *A (k,m,n)-selector is $(m-1,k-m)$-disjunct.*

Proof. For any k-set K of columns in a (k,m,n)-selector, the union of any $k-m+1$ columns is not contained in the union of the other $m-1$ columns since the $k-m+1$ columns contain at least one isolated column in K. □

Let $t_s(k,m,n)$ denote the minimum number of rows for a (k,m,n)-selector. De Bonis, Gasieniec and Vaccaro proved

Theorem 2.6.6 *For $1 \le m \le k < n$,*

$$t_s(k,m,n) < \frac{ek^2}{k-m+1} \ln \frac{n}{k} + \frac{ek(2k-1)}{k-m+1}.$$

Before proving Theorem 2.6.6, we first quote a lemma of Lovász [19] on hypergraph.

Lemma 2.6.7 *Consider a hypergraph H with vertex-set V and edge-set E. Let Δ be the maximum degree of a vertex and VC a vertex-cover of E. Then*

$$|VC| < \frac{|V|}{\min_{e \in E} |e|}(1 + \ln \Delta).$$

Proof of Theorem 2.6.6. We first transform a (k,m,n)-selector into a hypergraph $H_r(V,E_r)$.

For convenience, assume k divides n. Let V be the set $[n], n/k$. Let M_r be the incidence matrix with V as rows and $[n]$ as columns. Let I_k denote the identity matrix of order k and a_i the ith row of I_k. For $S = \{i_1, ..., i_k\} \subseteq \{1, ..., n\}$, let $M_V(S)$ be obtained from M_V by restricting to S. For $A \subseteq \{a_1, ..., a_k\}$, define edge

$$e_{A,S} = \{v \in V \mid \text{ row } v \text{ in } M_V(S) \text{ belongs to } A\}.$$

Let

$$E_w = \{e_{A,S} \mid |A| = w, S \subseteq \{1, ..., n\}\}.$$

Then any vertex-cover VC of H_{k-m+1} is a (k, m, n)-selector.

To see this, we represent VC as a $|VC| \times n$ binary matrix where the rows are the vertices in VC. For a fixed k-set S, let $VC(S)$ be obtained from VC by taking only columns in S. Suppose to the contrary that $VC(S)$ contains only a set A' of at most $m-1$ distinct isolated rows. Let A be any $(k-m+1)$-subset of $\{a_1, ..., a_k\} \setminus A'$. Then $e_{A,S}$ is an edge in H_{k-m+1}. But

$$e_{A,S} \cap VC = \emptyset,$$

contradicting the assumption that VC is a vertex-cover.

To apply Lemma 2.6.7, we compute the various parameters in H_{k-m+1}:

(i) $|V| = \binom{n}{n/k}$.

(ii) $|e_{A,S}| = (k-m+1)\binom{n-k}{n/k-1}$, since there are $k-m+1$ ways to choose i from $i \in A$, and with i fixed, the other $n/k-1$ 1-entries must be disjoint from the k indices in S.

(iii) $\Delta = \binom{k-1}{k-m}\binom{n/k}{1}\binom{n-n/k}{k-1}$, since for vertex v to be in $e_{A,S}$, S must choose one element from the 1-entries and $k-1$ elements from the 0-entries. For a fixed S, besides v is in A, the other $k-m$ elements or A must be chosen from the other $k-1$ elements of S.

By noting

$$\begin{aligned}
\frac{\binom{n}{n/k}}{\binom{n-k}{n/k-1}} &\leq k(\frac{n-k+1}{n-k-n/k+2})^{k-1} = k(1+\frac{n-k}{k(n-k+1)-(n-k)})^{k-1} \\
&\leq k(1+\frac{1}{k-1})^k < ek
\end{aligned}$$

and

$$\begin{aligned}
\binom{k-1}{k-m}\binom{n/k}{1}\binom{n-n/k}{k-1} &\leq e^{k-m}(\frac{k-1}{k-m})^{k-m}(\frac{n}{k})e^{k-1}(\frac{n-n/k}{k-1})^{k-1} \\
&= e^{2k-m-1}(1+\frac{m-1}{k-m})^{k-m}(\frac{m}{k})^k \\
&\leq e^{2k-1}(\frac{n}{k})^k,
\end{aligned}$$

we substitute (i), (ii) and (iii) into Lemma 2.6.7 and obtain Theorem 2.6.6. □

Chen, Hwang and Li [3] established a relation between separable matrices and (d, r)-disjunct matrices. For a binary matrix M and a set D of positive items. Let T_D denote the set of positive pools and $t_D = |T_D|$, N_D the set of columns contained in T_D and $n_D = |N_D|$. They proved

Lemma 2.6.8 *For a d-separable matrix,* $n_D \leq 2^{\lfloor t_D/d \rfloor} + d - 1$.

Proof. Let $D = \{D_1, ..., D_d\}$. Define

$$D_i^* = D_i \setminus \cup_{j \neq i} D_j \text{ for all } 1 \leq i \leq d.$$

Let C and C' be two columns in $N_D \setminus D$. Since $U(D)$ intersects every row in T_D,

$$\begin{aligned} C \setminus D_i^* &\subseteq \cup_{j \neq i} D_j, \\ C' \setminus D_i^* &\subseteq \cup_{j \neq i} D_j. \end{aligned}$$

By d-separability,

$$C \cup (\cup_{j \neq i} D_j) \neq C' \cup (\cup_{j \neq i} D_j).$$

Hence

$$C \cap D_i^* \neq C' \cap D_i^*,$$

i.e., all columns in N_D intersect D_i^* in different subsets. Let D_i^* have the minimum cardinality among all D_y^*. Then $|D_i^*| \leq \lfloor t_D/d \rfloor$. Hence D_i^* has only $2^{\lfloor t_D/d \rfloor}$ distinct subsets. Further, one must be subtracted since the intersection cannot be D_i^* or the union of that column with $\cup_{j \neq i} D_j$ yield the same outcome vector as D, violating the d-separability. Adding D back, we obtain Theorem 2.6.8. □

Define

$$t_d = \max_{|D|=d} t_D.$$

Theorem 2.6.9 *A d-separable matrix is* $(d, 2^{\lfloor t_d/d \rfloor} + d - 1)$*-disjunct.*

They also observed that for a $\bar{d}$-separable matrix, another 1 can be deleted from n_D since the intersecting subset cannot be the empty set too or the union of that column with $\cup_{j \neq i} D_j$ would be equal to $\cup_{j \neq i} D_j$, violating the $\bar{d}$-separability. Hence

Theorem 2.6.10 *A $\bar{d}$-separable matrix is* $(d, 2^{\lfloor t_d/d \rfloor} + d - 2)$*-disjunct.*

Example. It can be easily verified that

$$M = \begin{pmatrix} 1 & & & & 1 & & & 1 & 1 & & & \\ 1 & & & & & 1 & 1 & & & 1 & & \\ & 1 & & & 1 & & 1 & & & & 1 & \\ & 1 & & & & 1 & & 1 & & & & 1 \\ & & 1 & & 1 & & & & & 1 & & 1 \\ & & 1 & & & 1 & & & 1 & & 1 & \\ & & & 1 & & & 1 & & 1 & & & 1 \\ & & & 1 & & & & 1 & & 1 & 1 & \end{pmatrix}$$

is an 8×12 $\bar{2}$-separable matrix. t_d is obtained by taking $D = \{C_5, C_6\}$. Then

$$T_D = \begin{pmatrix} 1 & & & 1 & \\ 1 & & & & 1 \\ & 1 & & 1 & \\ & 1 & & & 1 \\ & & 1 & 1 & \\ & & 1 & & 1 \end{pmatrix}$$

with $t_d = 6$. By Lemma 2.6.8,

$$n_d = n_D \leq 2^{6/2} + 2 - 2 = 8,$$

while the actual n_D is 5.

2.7 Error-Tolerance

The *Hamming distance* of two column vectors is the number of different components between them, i.e., one 0 and the other 1. For a d-separable matrix, the Hamming distance of any two unions of d columns is at least one. Suppose that the Hamming distance of any two unions of d columns in a d-separable matrix is at least z. Then decoding d positive items from testing outcomes is still possible even if there exist up to $\lfloor (z-1)/2 \rfloor$ error-tests. Indeed, we can choose the d columns whose union is within Hamming distance $\lfloor (z-1)/2 \rfloor$ from the test-outcome vector. Such a matrix will be called *$(d; z)$-separable*. Similarly, we can define $(\bar{d}; z)$-separable. Clearly, $(\bar{d}, z)$-separable implies $(d; z)$-separable.

The following are basic properties of $(d; z)$-separable matrices.

Lemma 2.7.1 *Deleting any k rows, $k < c$, from a $(d; z)$-separable matrix results in a $(d; z-k)$-separable matrix.*

Proof. Two unions of d columns previously have distance z away is now at least distance $z - k$ away. □

Corollary 2.7.2 *The union of any d columns in a $(d; z)$-separable matrix has weight at least $d + z - 2$.*

Proof. Delete $z - 1$ rows where the union of the d columns is 1. Then the reduced matrix is still d-separable. By Corollary 2.1.4, it has weight at least $d - 1$. Hence the original weight is at least $z - 1 + d - 1 = z + d - 2$. □

D'yachkov, Rykov and Rashad [8] and Macula [20] gave a disjunct version of the $(d; z)$-separable matrix, which was further extended to the $(d, r; z)$-disjunct matrix by De Bonis, Gasieniec and Vaccaro [4]. A binary matrix is $(d, r; z)$-disjunct if the union of any r columns has at least z elements not in the union of any d other columns. Thus $(d, 1; z)$-disjunct is the $(d; z)$-disjunct of D'yachkov, Rykov and Rashad, $(d, r; 1)$-disjunct is (d, r)-disjunct studied in Section 2.6, and $(d, 1; 1)$-disjunct is just d-disjunct.

Lemma 2.7.3 *$(d; z)$-disjunct implies $(\bar{d}; z)$-separable.*

Proof. Suppose M is not $(\bar{d}; z)$-separable, i.e., there exist two sets D, D' of columns, $|D| \leq d$, $|D'| \leq d$, such that $U(D)$ and $U(D')$ differ in at most $z-1$ elements. Without loss of generality, assume $D \not\subseteq D'$. Then there exists a column $C \in D \setminus D'$ such that C and $U(D')$ differ in at most $z-1$ elements. Hence M is not $(d; z)$-disjunct. □

It follows from Lemma 2.7.3 that $(d; z)$-disjunct implies $(d; z)$-separable. But a stronger result is available.

Lemma 2.7.4 *$(d; z)$-disjunct implies $(d; 2z)$-separable.*

Proof. Suppose M is $(d; z)$-disjunct. Let D and D' be two d-sets of columns. Then $C \in D \setminus D'$ has at least z elements not in $U(D')$, and $C' \in D' \setminus D$ also has at least z elements not in $U(D)$. Hence $U(D)$ and $U(D')$ differ in at least $2z$ elements. □

By mimicking the proofs in Lemmas 2.2.7 and 2.2.8, we obtain

Lemma 2.7.5 *M is $(\bar{d}; z)$-separable if and only if it is $(d; z)$-separable and $(d-1; z)$-disjunct.*

Also, by mimicking the proof of Theorem 2.2.9, but noting a maximum of c rows is needed to remedy a set of $d+1$ columns not satisfying the $(d; c)$-disjunct condition, we obtain

Theorem 2.7.6 *Let M be a $(2d; z)$-separable matrix. Then we can obtain a $(d; z)$-disjunct matrix by adding at most z rows to M.*

By applying Lemma 2.7.4, we have

Corollary 2.7.7 *A $(2d; z)$-separable matrix becomes a $(d; 2z)$-separable matrix by adding at most z rows.*

For given d and z, D'yachkov, Rykov and Rashad [8] and Wei [26] studied $\lim(\log n/t)$ as $n \to \infty$ among other things, they obtain

Theorem 2.7.8 $t(d, n; z) \geq c\left[\frac{d^2 \log n}{\log d} + (z-1)d\right]$ *where c is a constant.*

On the upper bound side, we have

Theorem 2.7.9 *A $(d+z-1)$-disjunct matrix without an isolated column is $(d;z)$-disjunct.*

Proof. Suppose M is not $(d;z)$-disjunct, i.e., there exist $d+1$ columns $C_0, C_1, ..., C_d$ such that $|C_0 \setminus \cup_{j=1}^{d} C_j| = s < z$. Since M does not have an isolated column, there exist s other columns whose union contains the s elements in $C_0 \setminus \cup_{j=1}^{d} C_j$. Therefore, C_0 is contained in a set of $d+s < d+z$ columns, or equivalently, M is not $(d+z-1)$-disjunct. □

However, this may not be a good bound as $(d+z-1)$-disjunct is too much stronger than $(d;z)$-disjunct. We need a better construction to improve the upper bound. A useful lemma towards that goal is

Lemma 2.7.10 *A matrix is $(d;z)$-disjunct if $\underline{w} \geq d\bar{\lambda} + z$.*

Corollary 2.7.11 *A $(t, dk+z, k)$-packing is $(d;z)$-disjunct.*

Balding and Torney [1] first gave the above two results for the $(d;z)$-separable property.

The decoding algorithm, in principle, is to select the unique set D such that the difference between $U(D)$ and the outcome vector does not exceed k. However, exactly how to find D is a tricky matter, even for the $(d;z)$-disjunct matrix.

Wu, Li, Wu and Huang [27] proposed a decoding method based on the following lemma.

Lemma 2.7.12 *Suppose testing is based on a $(d;z)$-disjunct matrix. If the number of error tests is not more than $z-1$, then for any positive item i and negative item j, the number of negative pools containing i is always smaller than the number of negative pools containing j.*

Proof. Suppose the number of negative pools containing i is m. Then these m pools must incur error. Therefore, there are at most $z-1-m$ error tests turning negative outcomes to positive outcomes. Moreover, we note that if no error exists, the number of negative pools containing j is at least z by the definition of $(d;z)$-disjunctness. Hence, the number of negative pools containing j is at least $z-(z-1-m) = m+1 > m$.
□

The decoding algorithm is to compute the number of negative pools intersecting each item and select the d items with the smallest such numbers to be the positive items. This algorithm runs in $O(nt)$ time. However, it doesn't work for the $(\bar{d}, n)$ space. Indeed, Theorem 2.7.12 holds also for the $(\bar{d}, n)$ space. However, we do not know how many smallest ones should be selected as positive.

The above decoding algorithm was first described in [14] for the $(d;z)$-disjunct matrix but $(z-1)$-error-correcting was wrongly claimed (the correct claim should be $\lfloor (z-1)/2 \rfloor$-error-correcting).

For the $(\bar{d},n)$ space, Wu, Huang, Huang and Li [28] showed that the decoding algorithm of [14] also worked for the $(d;2k+1)$-separable matrix as long as it is d-disjunct.

Note that a $(d;z)$-disjunct matrix is d-disjunct and $(d;2z)$-separable. However, the inverse is not true. Indeed, there exists a $(\bar{2},5;4)$-separable and 2-disjunct matrix, which is not $(2;2)$-disjunct ($\delta^*(m,2,5)$ in [5, 14]).

An interesting question is whether the time complexity of decoding for $(d;2k+1)$-separable and d-disjunct matrices can be polynomial with respect to both d and n matrix. Wu *et al.* gave a positive answer to the question.

Theorem 2.7.13 *There exists a decoding algorithm for $(d;2k+1)$-separable and d-disjunct matrices, running in time $O((n+t)t^k)$.*

Proof. Let M be a $(d;2k+1)$-separable and d-disjunct $t \times n$ matrix. Given outcomes from t tests on a sample of n clones with at most d positive ones, assume that the number of errors is at most k. We describe a decoding approach as follows.

Let F be a set of pools with $|F| \leq k$. For each such F, we remove all clones in negative pools not in F and all clones in positive pools in F. If the number of remaining clones is at most d, then we compute the Hamming distance between the given test outcomes and the outcomes with all remaining clones as positives. If this Hamming distance does not exceed k, then we accept the result that all remaining ones are positive and the removed ones are negative. Clearly, this method runs in time $O((n+t)t^k)$.

Note that when F is indeed the error set, all negative clones will be removed by their presence in negative pools. Hence the remaining clones, at most d, constitute the positive set. Further, from the $(d;2k+1)$-separability of M, there exists exactly one sample with at most d positive clones such that the Hamming distance between the given test outcomes and the true outcomes on the sample does not exceed k. Since different F corresponds to different samples, only one F can be accepted. □

Corollary 2.7.14 *If the matrix is $(d;2k+1)$-separable and (d,r)-disjunct, then the checking would be whether the number of remaining items is at most $d+r$. The algorithm leaves a unique set of at most $d+r$ clones, including all positive ones.*

Corollary 2.7.15 *We can use a $(d,r;2k+1)$-disjunct matrix in Corollary 2.7.14.*

Is there a decoding method for $(d;2k+1)$-separable and d-disjunct matrices running in polynomial time with respect to n, d, t and k? Wu *et al.* gave a negative answer.

Theorem 2.7.16 *There does not exist a decoding algorithm for $(d; 2k+1)$-separable and d-disjunct matrix running in polynomial time with respect to n, d, t, and k unless NP=P.*

To prove this theorem, let us first study the decoding for $(\bar{d}; 2k+1)$-separable matrices.

The decoding for $(\bar{d}; 2k+1)$-separable matrices looks for a subset of at most d items such that the number of positive pools not hit by the subset plus the number of negative pools hit by the subset does not exceed k. Thus, this decoding problem is closely related to the following problem.

> HITTING AND NOT-HITTING (HNH) PROBLEM: Given two collections $\mathcal{C}$ and $\mathcal{D}$ of subsets of X and a positive integer d, find a subset A of at most d elements of X to minimize the total number of subsets in $\mathcal{C}$ not hit by A and of subsets in $\mathcal{D}$ hit by A.

Indeed, $\mathcal{C}$ is the collection of positive pools and $\mathcal{D}$ is the collection of negative pools. What we minimize is the number of error tests for the "hitting" subset over all possible set of positive clones. The difference is that for the decoding problem, we know that the minimum value of the objective function is at most k and want to find the subset to achieve this value.

Similarly, the decoding for $(d; 2k+1)$-separable and d-disjunct matrix is closely related to a variation of HNH problem. Note that by d-disjunctness, the set of positive clones is the complement to the union of negative pools. Thus, the minimization should be over all subsets A of cardinality at most d, satisfying the following property:

(*) $A = X - \cup_{B \in \mathcal{C} \cup \mathcal{D}, A \cap B = \emptyset} B$.

Formally, we state this variation as follows.

> HNH-(*) PROBLEM: Given two collections $\mathcal{C}$ and $\mathcal{D}$ of subsets of X and a positive integer d, find a subset A of at most d clones, satisfying (*), to minimize the total number of subsets in $\mathcal{C}$ not hit by A and subsets in $\mathcal{D}$ hit by A.

Lemma 2.7.17 *If decoding for $(d; 2k+1)$-separable and d-disjunct matrices can be done in polynomial-time with respect to n, t, d, and k, then the HNH-(*) problem can be solved in polynomial time.*

Proof. The decoding for $(d; 2k+1)$-separable and d-disjunct matrix is equivalent to the problem that knowing the minimum value of HNH-(*) problem is at most k, find a subset A of at most d clones, satisfying (*), to achieve the minimum. If there exists an algorithm K solving this problem in polynomial time with respect to n, t, d and k, then we may solve the HNH-(*) problem by applying this algorithm K repeatedly for $k = 1, 2, ..., t$. For each $k = 1, 2, ..., t$, if the algorithm K cannot find in polynomial time a subset A of at most d clones, satisfying (*), to achieve the k-value, then restart the algorithm for the next k-value, until a k-value is achieved by a subset A of at most d clones, satisfying (*). This clearly still runs in polynomial time. □

Next, we show that the HNH-(*) problem is NP-hard. To do so, we first study a variation of the vertex-cover problem as follows:

> VERTEX-COVER: Given a graph G with n vertices and a positive integer h, $0 < h \leq n$, determine whether G has a vertex-cover of size h.
>
> VERTEX-COVER-(*): Given a graph G and two positive integers m and h, determine whether G has a vertex subset of size at most h, covering at least m edges, such that the complement of the vertex-cover has no isolated vertex.

Lemma 2.7.18 VERTEX-COVER-(*) *is NP-complete.*

Proof. VERTEX-COVER-(*) is clearly in NP since we can guess the subset and verify in polynomial-time. We next construct a polynomial-time reduction from the well-known NP-complete problem VERTEX-COVER to VERTEX-COVER-(*).

Consider a graph G with n vertices and a positive integer h, $0 < h \leq n$. Let m be h plus the number of edges of G. We construct another graph G' from G by adding $h+1$ new vertices $v_0, v_1, ..., v_h$ and connecting v_0 to all vertices of G and $v_1, ..., v_h$ (Fig. 2.1). Next, we show that G has a vertex-cover of size h if and only if G has a

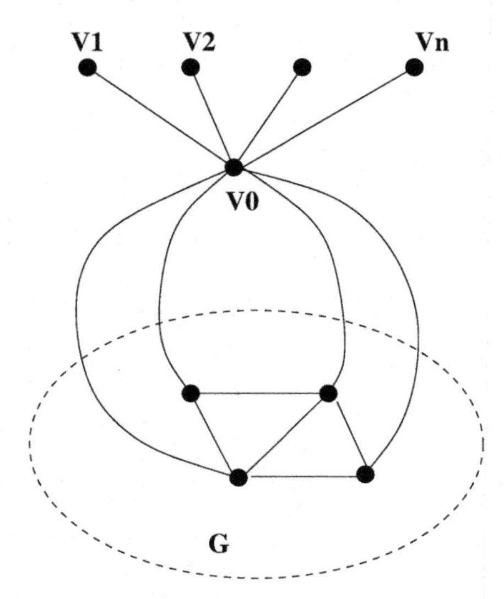

Figure 2.1: Construction of G' from G.

vertex subset of size h, covering at least m edges, such that the complement of the subset has no isolated vertex.

First, if G has a vertex-cover of size at most h, then this vertex cover in G' has the required property.

Next, assume that G' has a vertex subset A of size at most h, covering at least m edges, such that the complement of A contains no isolated vertex. Note that v_0 cannot be in A. In fact, if v_0 is in A, then each v_i for $i = 1, ..., h$ has to be in A; otherwise, v_i would be isolated in the complement. However, all $v_0, v_1, ..., v_h$ being in A contradicts the size of A. Since v_0 is not in A and each vertex other than v_0 covers exactly one edge not in G, A covers at least $m - h$ edges in G. Hence, A covers all edges of G. Therefore, all vertices of G in A is a vertex-cover of size at most h. If this vertex-cover has size smaller than h, then we can simply add in more vertices to achieve the size h. □

Finally, we finish the proof of Theorem 2.7.16 by proving the following lemma.

Lemma 2.7.19 *The HNH-(*) problem is NP-hard.*

Proof. Consider the decision version of the HNH-(*) problem:

> Decision version of the HNH-(*) problem: Given two collections $\mathcal{C}$ and $\mathcal{D}$ of subsets of X and two positive integers d and k, determine whether or not there exists a subset A of at most d elements, satisfying condition (*), and the total number of subsets in $\mathcal{C}$ not hit by A and subsets in $\mathcal{D}$ hit by B is at most k.

Now, we construct a polynomial-time reduction from VERTEX-COVER-(*) to the decision version of the HNH-(*) problem. Consider an instance of VERTEX-COVER-(*), consisting of a graph G and two positive integers m and h. Let X be the vertex set of G and $\mathcal{C}$ the edge set of G. Set $\mathcal{D} = \emptyset$, $k = |\mathcal{C}| - m$, and $d = h$. We show that G has a vertex subset H of size at most h, hitting at least m edges, such that its complement contains no isolated vertex if and only if X has a subset A of size at most d satisfying condition (*) and the number of subsets in $\mathcal{C}$ not hit by A is at most k.

If G has such a vertex subset H, then set $A = H$ which is a subset of X, with cardinality at most d and satisfying condition (*). Conversely, if X has such a subset A, then set $H = A$ and H is a required vertex subset for G. □

Every $(d; k+1)$-disjunct matrix is $(d; 2k+1)$-separable. However, for $\bar{d}$-separability, only a weaker result exists; every $(d; k+1)$-disjunct matrix is $(\bar{d}, k+1)$-separable. Could this result be improved? This is an open problem.

References

[1] D. J. Balding and D. C. Torney, Optimal pool designs with error detection, *J. Combin. Thy., Series A*, 74 (1996) 131-140.

[2] H. B. Chen and F. K. Hwang, Exploring the missing link among d-separable, $\bar{d}$-separable, and d-disjunct matrices, preprint, 2003.

[3] H. B. Chen, F. K. Hwang and C. M. Li, Bounding the number of columns which appear only in positive pools, *Taiwan. J. Math.*, to appear.

[4] A. De Bonis, L. Gasieniec and U. Vaccaro, Optimal two-stage algorithms for group testing problems, *SIAM J. Comput.*, 34 (2005) 1253-1270.

[5] D.-Z. Du and F. K. Hwang, *Combinatorial Group Testing and Its Applications (2nd Ed.)*, (World Scientific, 2000), pp. 167-168.

[6] A. G. Dyachkov and V. V. Rykov, Bounds of the length of disjunct codes, *Problems Control Inform. Thy.*, 11 (1982), 7-13.

[7] A. G. Dyachkov and V. V. Rykov, A survey of superimposed code theory, *Problems. Control Inform. Thy.*, 12 (1983) 1-13.

[8] A. G. D'yachkov, V. V. Rykov and A. M. Rachad, Superimposed distance codes, *Problems Contrd Inform. Thy.*, 12 (1983) 1-13.

[9] P. Erdös, P. Frankl and Z. Füredi, Families of finite sets in which no set is covered by the union of two others, *J. Combin. Thy.* A33 (1982) 158-166.

[10] P. Erdös, P. Frankl and D. Füredi, Families of finite sets in which no set is covered by the union of r others, *Israel J. Math.*, 51 (1985) 79-89.

[11] P. Frankl, On Sperner families satisfying an additional condition, *J. Combin. Thy.* A24 (1976) 1-11.

[12] Z. Füredi, On r-cover-free families, *J. Combin. Thy.*, A73 (1996) 172-173.

[13] J. H. Huang and F. K. Hwang, when is individual testing optimal for nonadaptive group testing? *SIAM J. Math.*, 14 (2001) 540-548.

[14] F. K. Hwang, On Macula's error-correcting pooling design, to appear in *Discrete Mathematics*, 268 (2003) 311-314.

[15] F. K. Hwang and V. T. Sós, Non-adaptive hypergeometric group testing, *Studia Scient. Math. Hungarica*, 22 (1987) 257-263.

[16] S. M. Johnson, A new upper bound for error correcting codes, *IEEE Trans. Inform. Thy.*, 8 (1962) 203-207.

[17] W. H. Kautz and R. R. Singleton, Nonrandom binary superimposed codes, *IEEE Trans. Inform. Thy.*, 10 (1964) 363-377.

[18] M. Li, Private communication.

[19] L. Lovász, On the ratio of optimal integral and fractional covers, *Disc. Math.*, 13 (1975) 383-390.

[20] A. J. Macula, Error-correcting nonadaptive group testing with d^e-disjunct matrices, *Disc. Appl. Math.*, 80 (1997) 217-222.

[21] Q. A. Nguyen and T. Zeisel, Bounds on constant weight binary superimposed codes, *Probl. Control & Inform. Thy.*, 17 (1988) 223-230.

[22] V. Rödl, On a packing and covering problem, *Euro. J. Combin.*, 5 (1984).

[23] M. Ruszinkó, On the upper bound of the size of the r-cover-free families, *J. Combin. Thy., Series A*, 66 (1994) 302-310.

[24] G. M. Saha and B. K. Sinha, Some combinatorial aspects of designs useful in group testing experiments, unpublished manuscript.

[25] E. Sperner, En Satz über Untermengen einer endlichen menge, *Math. Z.*, 27 (1928) 544-548.

[26] R. Wei, On cover-free families, preprint, 2004.

[27] W. Wu, C. Li, X. Wu and X. Huang, Decoding in pooling designs, *Journal of Combinatorial Optimization*, 7 (2003).

[28] W. Wu, Y. Huang, X. Huang and Y. Li, On error-tolerant DNA screening, *Discrete Applied Mathematics*, to appear.

3

Deterministic Designs

In this chapter we study the more traditional constructions of pooling designs usually of combinatorial nature, while the constructions of the following chapters are more algebraic and geometrical. There are many particular methods of constructions, but only of two general categories. One is by controlling the number of intersections of any pair of columns, and the other by transforming a q-ary matrix with certain properties to binary. We will refer to them as the *column-intersection method* and the *q-ary method*, respectively. Then we introduce some constructions for the special case $d = 2$.

3.1 t-Designs and t-Packing

The most natural combinatorial object tailored for column-intersection method is the block design since its incidence matrix with items as columns and blocks as rows possesses the property that every pair of columns intersect exactly λ times. Since each item appears in exactly r blocks, we have

Theorem 3.1.1 *The incidence matrix of a (v, b, r, k, λ)-design with items as columns and blocks as rows is $(d; z)$-disjunct with $d = \lfloor (r-1)/\lambda \rfloor$ and $z = r - d\lambda$.*

The only problem with this construction is $b \geq v$ by Fisher's inequality. Namely, there are more tests than items; so we might as well use the identity matrix I_n.

The Fisher's inequality can be avoided by relaxing the condition on exactly λ intersections to at most $\bar{\lambda}$ intersections. Note that the decrease of λ in some pairs of columns can only help the d-disjunct property since d is inversely proportional to λ.

One way to do it is to use the incidence matrix of a Steiner t-design with blocks as columns. Since each t-subset appears only in one block, two columns can intersect at most $t - 1$ times. Hence

Theorem 3.1.2 *The incidence matrix of a Steiner (t, v, k)-design with blocks as columns is $(d; z)$-disjunct with $d = \lfloor (k-1)/(t-1) \rfloor$ and $z = k - d(t-1)$.*

We can replace the Steiner (t, v, k)-design with a (t, v, k)-packing.

Corollary 3.1.3 *The incidence matrix of a (t, v, k)-packing with blocks as columns is $(d; z)$-disjunct with $d = \lfloor (k-1)/(t-1) \rfloor$ and $z = k - d(t-1)$.*

The *packing number* $PN(t, v, k)$ is the size of a maximum (t, v, k)-packing. Schönheim [20] obtained the following upper bound

$$PN(t, v, k) \leq \left\lfloor \frac{v}{k} \left\lfloor \frac{v-1}{k-1} \cdots \left\lfloor \frac{v-t+1}{k-t+1} \lambda \right\rfloor \cdots \right\rfloor \right\rfloor .$$

$PN(t, v, k)$ also has a close relation with coding theory. The following result can be found in [18] that

$$PN(v, k, t) = A(v, 2k - 2t + 2, k)$$

where $A(n, d, w)$ denotes the size of a maximum constant w-weight binary (n, d)-code.

Rödl [17] proved

Corollary 3.1.4 $PN(t, v, k) = (1 + o(1))\binom{v}{t}/\binom{k}{t}$.

A (t, v, k)-packing can also be constructed from some known combinatorial objects. The notion of block design has been extended to the *partially balanced incomplete block design* (PBIBD) by allowing λ to vary among the $\binom{t}{2}$ pairs of items within a certain structure. If there are m different values of λ, say, $\{\lambda_1 > \lambda_2 > \cdots > \lambda_m\}$, then the PBIBD is known as an m-associate PBIBD.

Theorem 3.1.5 *The incidence matrix of an m-associate PBIBD with blocks as rows is $(d; z)$-disjunct with $d = \lfloor (r-1)/\lambda_1 \rfloor$ and $z = r - d\lambda_1$.*

Saha, Pesotan and Raktoc [19] proved the original version of Theorem 3.1.5 for d-disjunctness.

The following is an example of using Theorem 3.1.5 to obtain a 2-disjunct matrix with fewer tests than the number of items.

Example 3.1 Let B be the 2-associated PBIBD consisting of the subsets:

$B_1 = \{1, 2, 3, 4\}$, $B_2 = \{5, 6, 7, 8\}$, $B_3 = \{9, 10, 11, 12\}$ $B_4 = \{13, 14, 15, 16\}$,

$B_5 = \{1, 5, 9, 13\}$, $B_6 = \{2, 6, 10, 14\}$, $B_7 = \{3, 7, 11, 15\}$, $B_8 = \{4, 8, 12, 16\}$,

$B_9 = \{1, 6, 11, 16\}$, $B_{10} = \{2, 7, 12, 13\}$, $B_{11} = \{3, 8, 9, 14\}$, $B_{12} = \{4, 5, 10, 15\}$.

It is easily verified that $\lambda_1 = 3$, $\lambda_{21} = 1$ and $\lambda_{22} = 0$. Since

$$\lambda_1 - d\lambda_{21} = 3 - 2 \cdot 1 = 1 > 0 ,$$

B is a 12×16 2-disjunct matrix.

A PBIBD is *group divisible* if the elements can be partitioned into groups of constant size such that an intergroup pair gets λ_{21} and an intragroup pair gets λ_{22}.

Theorem 3.1.6 *A group divisible PBIBD with groups of size* $g-1$, $\lambda_{21}=1$, $\lambda_{22}=0$ *yields a* $(g-1)$*-disjunct matrix.*

Proof. Take the groups and the blocks as columns. Add a row with r 1-entries for columns corresponding to groups and 0-entries for columns corresponding blocks. Then M', the enhenced M, has $\bar{\lambda}=1$ and $w=g$. By Lemma 2.3.9, M' is $(g-1)$-disjunct.

Example 3.2 Let B be the group divisible triple design consisting of eight blocks: $B_1=(1,3,5)$, $B_2=(2,4,6)$, $B_3=(2,3,7)$, $B_4=(1,4,8)$, $B_5=(1,6,7)$, $B_6=(2,5,8)$, $B_7=(4,6,8)$, $B_8=(4,5,7)$ with group $(1,2)$, $(3,4)$, $(5,6)$, $(7,8)$. Then M is a 9×12 2-disjunct matrix:

$$M=\begin{pmatrix} 1&0&0&0&1&0&0&1&1&0&0&0\\ 1&0&0&0&0&1&1&0&0&1&0&0\\ 0&1&0&0&1&0&1&0&0&0&1&0\\ 0&1&0&0&0&1&0&1&0&0&0&1\\ 0&0&1&0&1&0&0&0&0&1&0&1\\ 0&0&1&0&0&1&0&0&1&0&1&0\\ 0&0&0&1&0&0&1&0&1&0&0&1\\ 0&0&0&1&0&0&0&1&0&1&1&0\\ 1&1&1&1&0&0&0&0&0&0&0&0 \end{pmatrix}$$

3.2 Direct Construction

Motivated by the column-intersection method, Hwang and Sós [10] gave the following direct construction (with some modification from the original text).

Let T be a set of t elements and $\binom{T}{k}$ consist of all k-subsets of T. (A k-subset is a subset of k elements.) Define $r=\lceil t/(16d^2)\rceil$, $k=4dr$ and $m=4r$. Choose C_1 arbitrarily from $\binom{T}{k}$. Delete from $\binom{T}{k}$ all members which intersect C_1 in at least m rows. Choose C_2 arbitrarily from the updated $\binom{T}{k}$. Again delete from the updated $\binom{T}{k}$ all members which intersect C_2 in at least m rows. Repeat this procedure until the updated $\binom{T}{k}$ is empty. Suppose $C_1,\cdots,C_n$ have been chosen this way.

Theorem 3.2.1 $C_1,\cdots,C_n$ *constitute a* $t\times n$ d*-disjunct matrix with* $n\geq(2/3)3^{t/16d^2}$.

Proof. By construction, any two columns can intersect in at most $m-1$ rows, i.e., $\bar{\lambda}\leq m-1$. Note that $\underline{w}=k$. Hence, $d\bar{\lambda}=4dr-d<k=4dr$. By Lemma 2.3.9, $C_1,\cdots,C_n$ constitute a d-disjunct matrix.

At the jth step the number of members in the updated T^k set which intersect C_j in at least m rows is at most

$$\sum_{i=m}^{k}\binom{k}{i}\binom{t-k}{k-i}.$$

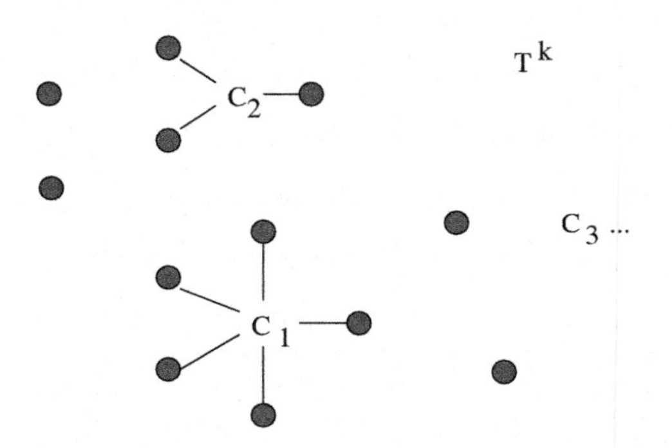

Figure 3.1: Choosing C_i from $\binom{T}{k}$.

Therefore,

$$n \geq \frac{\binom{t}{k}}{\sum_{i=m}^{k} \binom{k}{i}\binom{t-k}{k-i}} .$$

Set

$$b_i = \binom{k}{i}\binom{t-k}{k-i} .$$

For $3r \leq i < k = 4dr$

$$\frac{b_i}{b_{i-1}} = \frac{(k-i)^2}{(i+1)(t-2k+i+1)} \leq \frac{(4dr-3r)^2}{3r(t-8dr+3r)} \leq \frac{(4d-3)^2}{3(16d^2-8d+3)} < \frac{1}{3} .$$

Hence

$$\begin{aligned}
\sum_{i=m}^{k} b_i &< \sum_{i=m}^{k} \left(\frac{1}{3}\right)^{i-3r} b_{3r} \\
&< b_{3r}\left(\frac{1}{3}\right)^{-3r} \frac{\left(\frac{1}{3}\right)^m}{1-\frac{1}{3}} \\
&= b_{3r}\frac{\left(\frac{1}{3}\right)^{r-1}}{2} .
\end{aligned}$$

Since

$$\binom{t}{k} = \sum_{i=0}^{k} b_i > b_{3r} ,$$

we have

$$n \geq \frac{\binom{t}{k}}{\sum_{i=m}^{k} b_i} > 2 \cdot 3^{r-1} \geq 2 \cdot 3^{\frac{t}{16d^2}-1} .$$

□

Corollary 3.2.2 $t(d,n) \leq 16d^2(1 - \log_3 2 + (\log_3 2)\log n)$.

This bound is not as good as the nonconstructive bound given in Chapter 2.

3.3 Explicit Construction of Selectors

A (k,m,n)-selector is an $(m-1,k-m)$-disjunct matrix. A construction of (k,m,n)-selectors has been showed in the proof of Theorem 2.6.6. In this section, we present another method given by Chlebus and Kowalsk [3].

Their construction is based on (k,k,n)-selectors constructed by Kautz and Singleton [12] and (ℓ,d,ϵ)-dispersers constructed by Ta-Shma, Umans and Zuckerman [23].

A (ℓ,d,ϵ)-*dispersers* is a bipartite graph $H = (V_1, V_2, E)$ satisfying the following two conditions:

(H1) Each $A \subseteq V_1$ with $|A| \geq \ell$ is adjacent to at least $(1-\epsilon)|V_2|$ vertices of V_2.

(H2) Every vertex in V_1 has degree exactly d.

Let M be an $s \times n$ (h,h,n)-selector and $H = (V_1, V_2, E)$ a (k,d,ϵ)-disperser. Suppose $V_1 = \{0,1,...,n-1\}$, $V_2 = \{0',1',...,(g-1)'\}$ and rows of M are labeled by $0,1,...,s-1$. Define an $sg \times n$ binary matrix M^* as follows: For $v \in V_1$ and $i = as+b$ where $a \geq 0$ and $0 \leq b < s$, cell (i,v) of M^* contains a 1-entry if and only if v is a neighbor of vertex a' in V_2 and cell (b,v) of M contains a 1-entry.

Theorem 3.3.1 *If* $(1-\epsilon)gh \geq kd$, *then* M^* *is a* (k,m,n)*-selector.*

Proof. For contradiction, suppose M^* is not a (k,m,n)-selector. Then there a k-set A of columns and a $(k-m+1)$-subset $C \subseteq A$ such that every column in C is not isolated in A.

We claim that for every neighbor w' of C, w' has more than h neighbors in A. In fact, for otherwise, suppose w' has at most h neighbors in A and w' is a neighbor of v in C. Let $N_A(w')$ be the set of those neighbors. Note that $v \in N_A(w')$. Since M is a (h,h,n)-selector, there exists a row b intersecting $N_A(w')$ with exactly one column which is v. Moreover, the $(ws+b)$-th row of M^* equals the intersection of the b-th row of M and the neighbor $N(w')$ of w'. Note that $N(w') \cap A = N_A(w')$. Therefore, the $(ws+b)$-th row of M^* intersects A with exactly one column which is v, contradicting the assumption about C.

Now, since H is $(k-r+1,d,\epsilon)$-disperser, C has at least $(1-\epsilon)g$ neighbors. Thus, the number of edges between those neighbors and A is more than $(1-\epsilon)gh$. However, each vertex in V_1 has degree d. This implies that the number of edges between A and V_2 cannot exceed $dk \leq (1-\epsilon)gh$, a contradiction. □

Corollary 3.3.2 *The explicit construction of Chlebus and Kowalski can be done with at most* $O(\frac{k^2}{k-m+1} \text{polylog } n)$ *rows.*

Proof. Ta-Shma, Umans and Zuckerman constructed (ℓ, d, ϵ)-dispersers $H = (V_1, V_2, E)$ with $|V_1| = n$ and $g = |V_2| = \Theta(\ell d/\delta)$, where $\delta = O(\log^3 n)$ and $d = O(\text{polylog } n)$. Choose $h = \frac{c\delta k}{k-m+1}$ where c is constant satisfying

$$(1-\epsilon)|V_2|h = (1-\epsilon)c\Theta(dk) \geq dk.$$

Note that the construction of Kautz and Singleton gives a (h, h, n)-selector with $s = O((h \log n)^2)$ rows. Therefore, the number of rows of M^* is

$$sg = O(\frac{k^2}{k-m+1} \text{polylog } n).$$

Actually, it is Indyk [11] who gave the first explicit construction of (k, m, n)-selectors. However, Indyk's construction works only for $m \leq 3k/4$. The construction of Chlebus and Kowalsk [3] is the first one for $m \geq 3k/4$. The construction in the proof of Theorem 2.6.6 is a later work.

3.4 Grid Designs

A very practical and much used design to control column intersection is the grid design. There is a simple design for $d = 1$. For n items, choose an $m \times m$ grid where $m \geq \sqrt{n}$. Put all items at grid points. Take each grid line as a test. The positive item can be identified at the intersection of two lines receiving positive test outcome.

This idea can also be extended to $d = 2$. However, one grid is not enough. In fact, if a and b are two positives not collinear, then there may exist two items c and e collinear with a and b. They cannot be identified because a test containing c or e also contains a or b and hence receives positive outcome. (See Fig. 3.2.) To solve this

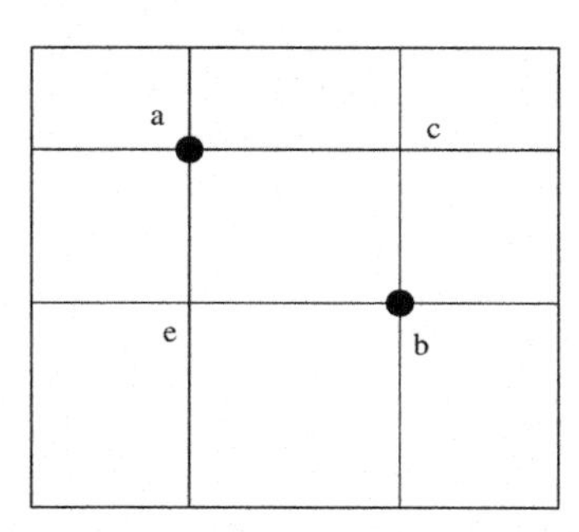

Figure 3.2: c and e are collinear with a and b.

problem, one can use two grids satisfying the following condition:

(C1) For any three items a, b and c, it occurs at most once that a is collinear with b on one line and also collinear with c on another line.

With condition (C1), if a negative item c cannot be identified in the first grid, then it must be able to be identified in the second grid. Therefore, for each grid, one may collect all intersections of positive tests to form a *suspect set*. Then intersection of the two suspect sets is the set of positives.

This idea also works for general d. To identify d positives, one needs $\binom{d}{2} + 1$ grids satisfying condition (C1) since an item can be in the intersection of two lines, each containing another positive item, at most $\binom{d}{2}$ times. All positives can be identified as intersection of the $\binom{d}{2} + 1$ suspect sets. Since each grid needs $2\lceil\sqrt{n}\rceil$ tests, the total number of tests is $O(d^2\sqrt{n})$.

One way to reduce the number of grids is to ask for a stronger condition between grids. Instead of (C1), Hwang [9] studied the following *unique collinearity condition*:

(C2) Every two items can be collinear in at most one grid.

The following theorem confirms that the number of grids is really reduced.

Theorem 3.4.1 $\lceil(d+1)/2\rceil$ *grids satisfying (C2) are enough to identify d positives (the corresponding incidence matrix is d-disjunct).*

Proof. Note that if a negative one cannot be identified in a grid, then it is collinear with at least two positives. By condition (C2), the existence of a negative one that is not identified by all $\lceil(d+1)/2\rceil$ grids would imply the existence of $2\lceil(d+1)/2\rceil > d$ positives, a contradiction. Therefore, the theorem is proved. □

Hwang also established the connection between grids satisfying (C2) and orthogonal Latin squares.

A *Latin square* of order k is a $k \times k$ array with entries in $\{1, 2, ..., k\}$ such that each element of $\{1, 2, ..., k\}$ appears exactly once in each row and exactly once in each column. Two Latin squares (a_{ij}) and (b_{ij}) of order k are said to be *orthogonal* if $\{(a_{ij}, b_{ij}) \mid 1 \le i, j \le k\} = \{(i, j) \mid 1 \le i, j \le k\}$. The following is an example of two orthogonal Latin squares:

$$\begin{pmatrix} 1 & 2 & 3 & 4 & 5 \\ 5 & 1 & 2 & 3 & 4 \\ 4 & 5 & 1 & 2 & 3 \\ 3 & 4 & 5 & 1 & 2 \\ 2 & 3 & 4 & 5 & 1 \end{pmatrix} \qquad \begin{pmatrix} 1 & 2 & 3 & 4 & 5 \\ 2 & 3 & 4 & 5 & 1 \\ 3 & 4 & 5 & 1 & 2 \\ 4 & 5 & 1 & 2 & 3 \\ 5 & 1 & 2 & 3 & 4 \end{pmatrix}$$

Theorem 3.4.2 *There exist g $k \times k$ grids satisfying (C2) if and only if there exist $2(g-1)$ orthogonal Latin squares of order k.*

Proof. If there exist $2(g-1)$ mutually orthogonal Latin squares, then one may divide them into $(g-1)$ pairs. For each pair, construct a grid by taking the entry of cell (i, j) in the first Latin square as the row index of item (i, j) in the grid and taking the entry of cell (i, j) in the second Latin square as the column index of item (i, j)

in the grid. One will obtain $g-1$ grids $F_1, ..., F_{g-1}$ together with grid F_g which has entry (i,j) in cell $((i,j))$. For example, the following grid is obtained from the two Latin squares given before.

$$\begin{pmatrix} 11 & 44 & 22 & 55 & 33 \\ 34 & 12 & 45 & 23 & 51 \\ 52 & 35 & 13 & 41 & 24 \\ 25 & 53 & 31 & 14 & 42 \\ 43 & 21 & 54 & 32 & 15 \end{pmatrix}$$

We now prove that no two different entries (x,y) and (x',y') can be collinear in two grids. Suppose to the contrary that F_u and F_v are two such grids, where F_u is constructed from (LS_{ur}, LS_{uc}), and F_v from (LS_{vr}, LS_{vc}). Without loss of generality, assume (x,y) and (x',y') are both in row i in F_u. Then entry i appears in both cells (x,y) and (x',y') in LS_{ur}. If (x,y) and (x',y') are both in row i' in F_v, then entry i' appears in both cells (x,y) and (x',y') in LS_{vr}, implying that LS_{ur} and LS_{vr} are not orthogonal. If (x,y) and (x',y') are both in column j' in F_v, then entry j' appears in both cells (x,y) and (x',y') in LS_{vc}, implying that LS_{uc} and LS_{vc} are not orthogonal. □

Hwang actually proved that the condition in Theorem 3.4.2 is also necessary.

The existence of orthogonal Latin squares is a mathematical problem well studied in combinatorial design. There do not exist two orthogonal Latin squares of order six. However, for odd prime power q, it is quite easy to construct $q-1$ mutually orthogonal Latin squares [16]. Based on those Latin squares, one can obtain $(q+1)/2$ grids satisfying (C2) as follows:

$$\begin{aligned} F_1 &= ((i,j)), \\ F_h &= ((s_{ij}, s_{ij}^h)), 2 \le h \le (q+1)/2, \end{aligned}$$

where

$$\begin{aligned} s_{ij} &\equiv i+j \pmod q \\ s_{ij}^h &\equiv (h-1)(j-i) \pmod q. \end{aligned}$$

Corollary 3.4.3 *If q is an odd prime power, then there are $(q+1)/2$ $q \times q$ grids satisfying (C2).*

Barillot *et al.* [2] generalized the idea to k-dimensional grids. Each line is viewed as a test while each grid point is a location to hold an item. A simple observation is that a $(d+1)$-dimensional grid can identify d positives. In fact, every grid point is the intersection of $d+1$ lines. Hence, for every negative item, there must exist a line through it which does not contain any positive item.

3.5 Error-Correcting Code

In a different direction Kautz and Singleton [12] searched among known families of conventional error-correcting codes for those which have desirable superimposition properties. They commented that binary codes do not lead to interesting superimposed codes (d-disjunct matrices). First of all, these codes include the zero-vector as a code word, so the corresponding matrix cannot be d-disjunct. The removal of the zero-vector does not solve the problem since the code usually contains a code word of large weight which contains at least one code word of smaller weight. Furthermore, if an error-correcting code of constant weight is extracted from an arbitrary error-correcting code, then Kautz and Singleton showed that the corresponding d-disjunct matrix will have very small values of d. So instead, they looked for codes based on q-nary error-correcting codes.

A q-nary error-correcting code is a code whose symbols are from the q-nary alphabet $\{0, 1, \cdots, q-1\}$. Kautz and Singleton constructed a binary superimposed code by replacing each q-nary symbol by a unique binary pattern. For example, such binary patterns can be the q-digit binary vectors with unit weight, i.e., the replacement is $0 \to 10\cdots0, 1 \to 010\cdots0, \cdots, q-1 \to 0\cdots01$. The *distance* l of a code is the minimum number of nonidentical symbols between two code words where the minimum is taken over all pairs of code words. Note that the distance l of the binary code is twice the distance l_q of the q-nary code it replaces, and the length t is q times the length t_q. Since the binary code has constant weight $w = t_q$, the corresponding $t \times n$ matrix is d-disjunct with

$$d \geq \left\lfloor \frac{w-1}{\lambda} \right\rfloor = \left\lfloor \frac{t_q - 1}{t_q - l_q} \right\rfloor$$

from Lemma 2.3.9.

To maximize d with given t_q and n_q, one seeks q-nary codes whose distance l_q is as large as possible. Kautz and Singleton suggested the class of *maximal-distance separable* (MDS) q-nary codes where the t_q symbols can be separated into k_q information symbols and $t_q - k_q$ check symbols. Singleton [22] showed that for an MDS code

$$l_q \leq t_q - k_q + 1 \ .$$

Thus for those MDS codes achieving this upper bound distance,

$$d \geq \left\lfloor \frac{t_q - 1}{k_q - 1} \right\rfloor ,$$

(a deeper analysis proves the equality). Also the k_q information symbols imply a total of $n_q = q^{k_q}$ code words.

Kautz and Singleton also commented that the most useful MDS q-nary codes for present purposes has q being an odd prime power satisfying

$$q + 1 \geq t_q \geq k_q + 1 \geq 3 \ .$$

Therefore for given d

$$q \geq t_q - 1 \simeq (k_q - 1)d ;$$

q is certainly too large for practical use unless k_q is very small, like two or three.

One may replace the q symbols by binary patterns more efficient than the aforementioned unit vector pattern, provided only that the q binary patterns for replacement form a d-disjunct matrix themselves. Clearly, the length of such binary patterns can be much shorter than the unit vector patterns.

Such a replacement can be regarded as a method of composition in which a small $t_0 \times n_0$ d_0-disjunct matrix is converted into a larger $t_1 \times n_1$ d_1-disjunct matrix through the medium of a t_q-digit q-nary code having k_q independent symbols where (q is a prime power satisfying $t_q - 1 \leq q \leq n_q$)

$$\begin{aligned} t_1 &= t_0 t_q , \\ n_1 &= q^{k_q} , \\ d_1 &= \min\left\{d_0, \left\lfloor \frac{t_q - 1}{k_q - 1} \right\rfloor\right\} . \end{aligned}$$

For the unit weight code, $t_0 = n_0 = d_0 = q$, i.e., the identity matrix of order q. Starting with a unit weight code and keeping d fixed, repeated compositions can be carried out to build up arbitrarily large d-disjunct matrices.

By setting $t_1 = qt_q$, Nguyen and Zeisel [15] used a result of Zinoviev [28] and the Johnson bound (Lemma 2.4.1) to prove

Theorem 3.5.1 *If $d < an^{1/k}$ for some integer $k > 2$ and constant a, then as $d \to \infty$, there exists a constant weight d-disjunct matrix such that*

$$t = (k-1)\frac{d^2 \log n}{\log d} \quad as\ n \to \infty .$$

Different q-nary codes can be used at each stage of the composition. If the same type of q-nary code is used except that q is replaced by $q = n_1$, then a second composition yields

$$\begin{aligned} t_2 &= t_0[1 + d(k_q - 1)]^2 \\ n_2 &= q^{(k_q)^2} \end{aligned}$$

where

$$d(k_q - 1) \leq q^{k_q} .$$

Kautz and Singleton also discussed the optimal number of compositions.

3.6 Transversal Designs

A pooling design is a *transversal* if it can be divided into disjoint families, each of which is a partition of all items. Hence, pools in each family are disjoint. The grid designs discussed in the last section are examples of transversal designs. Transversal designs are used very frequently in practice because implementation is easy and performance is quite good.

Here, we introduce an interesting relation between transversal designs and integer matrices given in [5].

For each transversal design, one may define a matrix with rows indexed by families and columns indexed by items and a cell (i, j) contains entry k if and only if item j belongs to the kth pool in the ith family.

For each $f \times n$ matrix, one may construct a transversal design with f families and n items as follows: Use entries on the ith row to index pools in the ith family. The pool with index k in the ith family contains the jth item if and only if cell (i, j) contains entry k in the matrix. For example, matrix

$$\begin{pmatrix} 1 & 1 & -1 & -1 \\ 2 & 3 & 2 & 3 \\ 0 & 1 & 1 & 2 \end{pmatrix}$$

represents the transversal design

$$\begin{matrix} 1 & 1 & 0 & 0 \\ 0 & 0 & 1 & 1 \\ 1 & 0 & 1 & 0 \\ 0 & 1 & 0 & 1 \\ 1 & 0 & 0 & 0 \\ 0 & 1 & 1 & 0 \\ 0 & 0 & 0 & 1 \end{matrix}$$

Clearly, each matrix yields a transversal design, but many can yield the same transversal design.

A matrix is called q-nary $(d, 1)$-disjunct if it is q-nary and for any column C, there does not exist a set D of d other columns such that every element of C appears in a column of D in the same row.

Theorem 3.6.1 *A $t \times n$ q-nary $(d, 1)$-disjunct matrix M yields a $t' \times n$ d-disjunct matrix with $t' \leq tq$.*

Proof. One may transform M to a binary matrix M' in the following way: Replacing each row R_i of M by several rows indexed with entries of R_i. For each entry x of R_i, the row with index x is obtained from R_i by turning all x's to 1-entries and all others to 0-entries. For example, row $(1, 1, -1, -1)$ is replaced by two rows $(1, 1, 0, 0)$ and

$(0, 0, 1, 1)$. Then it is easy to verify that the binary matrix M' is d-disjunct if and only if M has the property stated in the theorem. □

It is worth mentioning that a d-disjunct binary matrix must be $(d, 1)$-disjunct, however, a $(d, 1)$-disjunct binary matrix may not be d-disjunct. The following is an example which is binary $(2, 1)$-disjunct, but not 2-disjunct.

$$\begin{pmatrix} 1 & 1 & 0 & 0 \\ 1 & 0 & 1 & 0 \\ 0 & 1 & 1 & 0 \end{pmatrix}$$

Du, Hwang, Wu and Znati [5] found an interesting way to construct q-nary $(d, 1)$-disjunct matrices as follows.

Consider a finite field $GF(q)$ of order q. Suppose k satisfies

$$n \le q^k \tag{3.6.1}$$

and

$$t = d(k-1) + 1 \le q. \tag{3.6.2}$$

Construct a $f \times n$ matrix $M(d, n, q, k)$ as follows: Its column indices are polynomials of degree $k - 1$ over the finite field $GF(q)$. Its row indices are t distinct elements of $GF(q)$. The cell (x, g) contains element $g(x)$ of $GF(q)$.

Theorem 3.6.2 *$M(d, n, q, k)$ is a q-nary $(d, 1)$-disjunct matrix.*

Proof. Suppose to the contrary that $M(d, n, q, k)$ is not $(d, 1)$-disjunct. Then it has a column g_0 contained in the union of other d columns $g_1, ..., g_d$. That is, for each row index x_i, $g_0(x_i) = g_j(x_i)$ for some $j \in \{1, ..., d\}$. Note that there are $d(k-1)+1$ rows. Thus, there exists a g_j $(1 \le j \le d)$ such that $g_0(x_i) = g_j(x_i)$ for at least k row indices x_i. It follows that $g_0 = g_j$, a contradiction. □

By (3.6.1) and (3.6.2), k and q should be chosen to satisfy

$$\log_q n \le k \le \frac{q-1}{d} + 1. \tag{3.6.3}$$

There exists a positive integer k satisfying (3.6.3) if q satisfies

$$\log_q n \le \frac{q-1}{d}. \tag{3.6.4}$$

That is, it is sufficient to choose q satisfying

$$n^d \le q^{q-1}. \tag{3.6.5}$$

Let q_0 be the smallest number q satisfying (3.6.5). Then, we have the following estimation on q_0.

Lemma 3.6.3

$$q_0 = (1 + o(1))\frac{d\log_2 n}{\log_2(d\log_2 n)}.$$

Moreover,

$$q_0 \le 1 + \frac{2d\log_2 n}{\log_2(d\log_2 n)}$$

for $n^d \ge 2^4$.

Proof. Set

$$q_1 = 1 + (1 + h(d,n))\frac{d\log_2 n}{\log_2(d\log_2 n)},$$

where

$$h(d,n) = \frac{\log_2\log_2(d\log_2 n)}{\log_2(d\log_2 n) - \log_2\log_2(d\log_2 n)}.$$

Note that $h(d,n) \ge 0$. Therefore,

$$\begin{aligned}(q_1 - 1)\log_2 q_1 &> (q_1 - 1)\log_2(q_1 - 1)\\ &\ge \frac{(1 + h(d,n))d\log_2 n}{\log_2(d\log_2 n)} \cdot \log_2\frac{(1 + h(d,n))d\log_2 n}{\log_2(d\log_2 n)}\\ &> d\log_2 n.\end{aligned}$$

That is, q_1 satisfies (3.6.5). It follows that $q_0 \le q_1$. Note that $h(d,n) = o(1)$. Hence,

$$q_0 = (1 + o(1))\frac{d\log_2 n}{\log_2(d\log_2 n)}.$$

Moreover, for $n^d \ge 2^4$, $d\log_2 n \ge 4$. Hence, $2^{d\log_2 n} \ge (d\log_2 n)^2$. Thus, $d\log_2 n \ge 2\log_2(d\log_2 n)$. It follows that $h(d,n) \le 1$. Therefore,

$$q_0 \le 1 + \frac{2d\log_2 n}{\log_2(d\log_2 n)}$$

for $n^d \ge 2^4$. □

We need to find a prime power q satisfying

$$q \ge q_0.$$

Then, we can choose

$$k = \lceil \log_q n \rceil.$$

For such a choice of k, we have

$$t = d(k - 1) + 1 \le d(\lceil \log_q n \rceil - 1) + 1 \le d(\lceil \log_{q_0} n \rceil - 1) + 1 \le q_0.$$

Since each family contains at most q pools, the total number of tests is at most $q_0 q$.

Theorem 3.6.4 *There exist a prime power q and a positive integer k, satisfying (3.6.1) and (3.6.2), such that $M(d,n,q,k)$ is $(d,1)$-disjunct design with at most $2q_0^2$ tests.*

Proof. Set $q = 2^{\lceil \log_2 q_0 \rceil}$. Then q is a prime power satisfying $q_0 \le q < 2q_0$. Therefore, $qq_0 < 2q_0^2$. □

Corollary 3.6.5 *There exists a $(d,1)$-disjunct design $M(d,n,q,k)$ with at most*

$$(2+o(1))(\frac{d\log_2 n}{\log_2(d\log_2 n)})^2$$

tests.

Du *et al.* [6] further reduced the number of tests with the following multiplication theorem about q-nary $(d,1)$-disjunct matrices.

Theorem 3.6.6 *If there exist a q-nary $(d,1)$-disjunct $t \times n$ matrix M_1 and a q'-nary $(d,1)$-disjunct $t' \times q$ matrix M_2, then there exists a q'-nary $(d,1)$-disjunct $tt' \times n$ matrix M_3.*

Proof. M_3 can be constructed from M_1 and M_2 by labeling columns of M_2 with $0,1,...,q-1$ and replacing each entry of M_1 by a corresponding column of M_2. Consider $d+1$ columns $C_0, C_1, ..., C_d$ of M_3. They are obtained from $d+1$ columns $C'_0, C'_1, ..., C'_d$ of M_1, respectively. Since M_1 is $(d,1)$-disjunct, C'_0 has at least one component a_0 not contained in the corresponding component $\{a_1, ..., a_{d'}\}$ $(d' \le d)$ of the union of $C'_1, ..., C'_d$. That is, a_0 is different from $a_1, ..., a_{d'}$. Hence, a_0 corresponds to a column C''_0 of M_2, which is different from those columns $C''_1, ..., C''_{d'}$ of M_2, corresponding to $a_1, ..., a_{d'}$, respectively. Since M_2 is $(d,1)$-disjunct, C''_0 is not contained in the union of $C''_1, ..., C''_{d'}$. Therefore, C_0 is not contained in the union of $C_1, ..., C_d$. □

Note that M_3 has $tt'q'$ tests. If $t'q' < q$, then $tt'q' < tq$, that is, the number of tests in M_3 is less than the number of tests in M_1. An observation made by Du *et al.* is that applying the multiplication theorem once result in a transversal design with nearly

$$c\frac{(d\log_2 n)\cdot(d\log_2(d\log_2 n))}{(\log_2(d\log_2(d\log_2 n)))^2}$$

tests, and applying the multiplication theorem ℓ times may result in a transversal design with nearly

$$c\frac{(d\log_2 n)\cdot\overbrace{(d\log_2\cdots(d\log_2 n))}^{1+\ell}}{(\log_2\underbrace{(d\log_2\cdots(d\log_2 n)}_{1+\ell}))^2}$$

tests, where c is a constant. As ℓ gets large enough, $\log_2(\underbrace{d\log_2\cdots(d\log_2}_{1+\ell} n))$ and $\log_2(\underbrace{d\log_2\cdots(d\log_2}_{\ell} n))$ are very close. Therefore, this number of tests reaches almost the lower bound

$$\frac{d^2\log_2 n}{\log_2 d}.$$

Du *et al.* also extended their construction to the error-tolerant case. They showed that a transversal design is $(d;e)$-disjunct if and only if its integer matrix representation has the property that every column has at least e components not contained in the union of some other d columns. An integer matrix satisfying this property is called a $(d,1;e)$-*disjunct matrix*.

Consider q and k satisfying

$$n \le q^k \tag{3.6.6}$$

and

$$t = d(k-1) + e \le q. \tag{3.6.7}$$

Construct a $t \times n$ matrix $M(d,n,q,k,e)$ as follows: Its column indices are polynomials of degree k over the finite field $GF(q)$. Its row indices are t distinct elements of $GF(q)$. The cell (x,g) contains element $g(x)$ of $GF(q)$.

Theorem 3.6.7 *$M(d,n,q,k,e)$ is $(d,1;e)$-disjunct transversal design.*

Proof. Suppose $M(d,n,q,k,e)$ is not $(d,1;e)$-disjunct. Then it has a column g_0 which has at least $t-e+1$ components contained in the union of some other d columns $g_1,\ldots,g_d$. Thus, there exists a column g_j containing at least k components of g_0. That is, for at least k row indices x_i, $g_0(x_i) = g_j(x_i)$. Therefore, $g_0 = g_j$, a contradiction. □

By an argument similar to the above, we can also obtain the following.

Theorem 3.6.8 *By properly choosing q and k, we can obtain a $M(d,n,q,k,e)$ with at most $2q_e^2$ tests where*

$$q_e = e + (2+o(1))\left(\frac{2d\log_2 n}{\log_2(d\log_2 n)}\right)^2.$$

For $(d,1;e)$-disjunct matrices, there also exists a multiplication theorem as follows:

Theorem 3.6.9 *If there exist a q-nary $(d,1;e)$-disjunct $t\times n$ matrix and a q'-nary $(d,1;e)$-disjunct $t'\times q$ matrix, then there exists a q'-nary $(d,1;e)$-disjunct $t't\times n$ matrix.*

This multiplication theorem can also be used to reduce the number of tests.

3.7 The $d = 2$ Case

There are some special constructions for the case $d = 2$.

Let H^T denote the transpose of the parity check matrix of a conventional binary d-error-correcting code. Then it is known (p. 33 [16]) that H^T has the property that the modulo-2 sums of any up to d columns are distinct. This property is exactly what is desired for $\bar{d}$-separable matrices except that the sum is Boolean sum for the latter. Kautz and Singleton suggested the following transformation for small d.

For $d = 2$, transform $0 \to 01$ and $1 \to 10$. The modulo-2 addition table before transformation and the Boolean addition table after transformation are given in the following:

$\oplus$	0	1
0	0	1
1	1	0

$\vee$	0	1
0	01	11
1	11	10

Thus if two pairs of columns have different modulo-2 sums, they also have different Boolean sums under this transformation. Namely, the 2-error correcting property is translated to the 2-separable property. Further, it is easily verified that after transformation, the Boolean sum of two columns can equal a third column only if the two columns are identical, contradicting the fact that a parity check matrix has no identical rows. Also, no column or the sum of two columns can equal the 0-column. Hence, the transformed matrix is $\bar{2}$-separable.

The parity check matrices of the family of Bose-Chaudhuri codes (p. 123 [16]) for $d = 2$ have $2^k - 1$ rows and no more than $2k$ columns for every $k \geq 2$. Hence they yield $\bar{2}$-separable matrices with $4k$ rows and $2^k - 1$ columns. We have

Theorem 3.7.1 $2 \log n \leq t_s(\bar{2}, n) \leq 4 \log(n + 1)$.

Note that the lower bound is from the fact that the $\binom{n}{2} + \binom{n}{1}$ distinct choices of D must not exceed the 2^t outcome vectors.

While bounds of $t_s(2, n)$ can be obtained from Theorem 3.7.1 by using Corollary 2.2.10, Lindstrom [13] gave a slightly different result with an independent proof of the upper bound.

Theorem 3.7.2 $2 \log(n - 1) - 1 \leq t_s(2, n) \leq 4 \log n$.

Proof. The lower bound is derived from the fact that there are $\binom{n}{2}$ pairs of columns whose unions must all be distinct t-vectors (there are 2^t of them).

To prove the upper bound, consider the set V of all vectors (x, x^3) with $x \in GF(2^t)$. Then V has cardinality 2^t. If

$$(x, x^3) + (y, y^3) = (u, v)$$

for two elements $x \neq y$ in $GF(2^t)$, then

$$x + y = u \neq 0,$$

and

$$-3xy = v/u - u^2 .$$

Substituting $y = u - x$ in the second equation and solve for x. Since an equation of the second degree cannot have more than two roots in the field, and since x' being a solution of x implies so is $u - x'$, the two solutions x_1 and x_2 of x must satisfy $x_1 + x_2 = u$, i.e., $\{x, y\}$ is uniquely determined by (u, v). Therefore V induces a $2t \times 2^t$ 2-separable matrix under modulo 2 arithmetic. A Boolean sum transformation doubles t. □

Kautz and Singleton also studied the construction of $\bar{2}$-separable matrices with constant weight two. Note that under constant weight, $\bar{2}$-separable is the same as 2-separable. Let M be such a matrix. Construct a simple graph $G(M)$ where vertices are the row indices and edges are the columns. Then $G(M)$ contains no 3-cycle or 4-cycle since such a cycle corresponds to two column-pairs with the same Boolean sum. Regular graphs (each vertex has the same degree r) of this type were studied by Hoffman and Singleton [7] with only four solutions:

$$r = 2 \quad t = 5 \quad n = 5$$

$$r = 3 \quad t = 10 \quad n = 15$$

$$r = 7 \quad t = 50 \quad n = 175$$

$$r = 57 \quad t = 3250 \quad n = 92625$$

which satisfy the conditions $t = 1 + r^2$ and $n = r(1 + r^2)/2$. Thus $n = t\sqrt{t-1}/2 \rightarrow t^{3/2}/2$ asymptotically. This result was also obtained by Vakil, Parnes and Raghavarao [25].

Relaxing the regularity requirement, one can use a $(v, k, b, r, 1)$ BIBD to construct a $(v + b) \times kb$ 2-separable matrix. Let x_{ij} denote the jth item in block i. Then each column is labeled by an x_{ij}, and each row by an item I or a block B. The column x_{ij} intersects row I if $x_{ij} = I$, and intersects row B if block i is B. For fixed t, n is maximized by using a symmetric BIBD where $v = b$ and $k = r$. From $\binom{v}{2} = v\binom{k}{2}$ or $v - 1 = k(k-1)$, it can then be shown that

$$n = \frac{t}{4}(1 + \sqrt{2t-3}) \rightarrow \frac{t^{3/2}}{2\sqrt{2}} .$$

Weideman and Raghavarao [27] proved a relation between $\bar{w}$ and $\bar{\lambda}$ for the $\bar{2}$-separable matrix. Huang and Hwang [8] observed that the relation also holds for the d-disjunct matrices.

Lemma 3.7.3 *Any d-separating matrix with $\bar{\lambda} = 1$ can be reduced to one with $\bar{w} = d + 1$.*

Proof. Replacing all columns of weight greater than $d+1$ by arbitrary $(d+1)$-subsets preserves the $\bar{2}$-separability since under the condition $\bar{\lambda} = 1$ no $(d+1)$-subset can be contained in the union of any d columns. The reduced matrix also preserves $\bar{\lambda}$.

Theorem 3.7.4 *Suppose that $\bar{\lambda} = 1$ in a $\bar{2}$-separable matrix. Then*

$$n \leq t(t+1)/6 \ .$$

Proof. From Lemma 3.7.3 assume $\bar{w} = 3$. A column of weight 2 can be represented by the pair of intersecting row indices. The 2-separable property imposes the condition that if columns $\{x, y\}$ and $\{x, z\}$ are in the design, then rows y and z cannot intersect. For if they intersect at column C, then

$$\{x, y\} \cup C = \{x, z\} \cup C \ .$$

Suppose that there are p columns of weight 2. These p pairs generate $2p$ row indices. There are at least $\max\{2p - t, 0\}$ pairs of indices sharing an index ("at least", because $(x, y), (x, z), (x, v)$ generate 3 such pairs). Therefore the p columns of weight 2 generate $p + \max\{2p - t, 0\}$ pairs of indices which cannot intersect in columns of weight exceeding 2. Thus the number of available pairs to intersect in columns of weight exceeding 2 is at most

$$\binom{t}{2} - p - \max\{2p - t, 0\} \leq \binom{t}{2} - 3p + t.$$

Since each column of weight exceeding 2 generates at least three intersecting pairs,

$$n \leq p + \left[\binom{t}{2} - 3p + t\right] /3 = t(t+1)/6.$$

□

Weideman and Raghavarao showed that the upper bound can be achieved for $t \equiv 0$ or $2 \pmod 6$ by taking $t/2$ pairs as groups and the $t(t-1)/6$ blocks of $S(2, 3, t)$ as columns. They also showed in a subsequent paper [26] that even for other values of t the upper bound can still often be achieved. Vakil and Parnes [24], by crucially using the constructions of group divisible triple designs of Colbourn, Hoffman and Rees [4], gave the following theorem.

Theorem 3.7.5 *The maximum n in a 2-separable matrix with $\bar{\lambda} = 1$ is*

$$n = \lfloor t(t+1)/6 \rfloor.$$

Since the constructions consist of many subcases, the reader is referred to [24] for details.

A variation of replacing the condition $\bar{\lambda} = 1$ by the weaker one $\bar{w} = 3$ was studied in [21, 24].

Kautz and Singleton tailored the composition method discussed in Section 3.3 to the $d = 2$ case. From a $t \times n$ 2-separable matrix M, construct a $(2t + 2p - 1) \times n^2$ 2-separable matrix M'. Column C_i of M' consists of three sections a_i, b_i, c_i, where $a_i \times b_i$, $1 \le i \le n^2$, enumerate all column pairs of M, and c_i, depending on a_i and b_i, has length $2p - 1$. The condition on c_i can be more easily seen by constructing a $t \times t$ matrix C whose rows (columns) are labeled by the columns of M chosen for $a_i(b_i)$ and the entries in the cell (a_i, b_i) yields c_i (a $(2p-1) \times 1$ vector). Denote the entry in cell (x, y) by C_{xy}. Then the condition, called the *minor diagonal condition,*, to be satisfied is

$$C_{xy} \vee C_{uv} \ne C_{xv} \vee C_{uy} ,$$

since otherwise the column-pair with first two sections $a_x b_y$, $a_u b_v$ and the column-pair with first two sections $a_x b_v$, $a_u b_y$ would have the same Boolean sum.

Starting with a 3×3 weight-one code and using the matrix

$$C = C_1 = \begin{pmatrix} 1 & 0 & 0 \\ 0 & 1 & 0 \\ 0 & 0 & 1 \end{pmatrix}$$

to construct the third section ($p = 1$ here), one obtains a 7×9 2-separable matrix. So the next matrix $C = C_2$ is of size 9×9. In general at the pth iteration the matrix C_p is of size $3^p \times 3^p$. Kautz and Singleton gave a method to construct C_p from C_{p-1} which preserves the minor diagonal condition where each entry in C_p is a $(2p-1) \times 1$ vector. Thus C_p yields a $t_p \times n_p$ 2-separable matrix with

$$t_p = 2t_{p-1} + 2p - 1 \text{ (or } t_p = 6 \cdot 2^p - 2p - 3)$$

and

$$n_p = 3^{2^p} .$$

It follows that $n \to 3^{t/6}$ asymptotically.

Similar composition can be used to grow a 2-disjunct matrix with parameters

$$t_p = 3\ t_{p-1}\ (or\ t_p = 3^{p+1})$$

and

$$n_p = 3^{2^p} .$$

Then $n \to 3^{t^{\log_3 2}/2}$.

Balding and Torney [1] solved $n(1, t; z)$ and $n(2, t; z)$. They observed the following lemma.

Lemma 3.7.6 *A $t \times n$ matrix M is (d,z)-disjunct if and only if $|C_i \backslash U(s)| \geq z$ for every set s of columns with $|s| \leq d$, and every $i \in \{1, ..., n\} \backslash s$.*

Proof. Necessity. Suppose $|C_i \backslash U(s)| \leq z-1$. Let $s' = C_i \cup s$. Then $U(s') = U(s)$ if the $z-1$ tests which intersect C_i but not s are all erroneous.

Sufficiency. Let s and s' be two sets of columns with $|s| \leq d$. If $s' \backslash s \neq \emptyset$, then there exists a column $C' \in s' \backslash s$ such that $|C' \backslash U(s)| > z-1$. If $s' \backslash s = \emptyset$, then $s \backslash s' \neq \emptyset$ and $|s'| \leq d$. Again, there exists a column $C \in s \backslash s'$ such that $|C \backslash U(s')| > z-1$. So in either case $U(s)$ and $U(s')$ differ at more than $z-1$ bits. □

Corollary 3.7.7 *M is $(d;z)$-disjunct if and only if $|C_i \backslash C_j| \geq z$ for all distinct columns C_i and C_j.*

Corollary 3.7.8 *If M is a $(v, dv+z, t)$-packing (every two columns intersect at most v times), then M is $(d;z)$-disjunct.*

Theorem 3.7.9 *A $(1;z)$-disjunct $t \times n$ matrix M satisfies*

$$n \leq \frac{1}{K_{z-1}} \binom{t}{\lfloor t/2 \rfloor},$$

where $K_0 = 1$ and for z odd,

$$K_{z-1} = \sum_{s=0}^{(z-1)/2} \binom{\lfloor t/2 \rfloor}{s} \binom{\lceil t/2 \rceil}{s},$$

while for z even,

$$K_{z-1} = K_{z-2} + \frac{1}{T} \binom{\lfloor t/2 \rfloor}{z/2} \binom{\lceil t/2 \rceil}{z/2},$$

with $T = \lfloor 2 \lfloor \frac{t}{2} \rfloor / z \rfloor$.

Proof. Lubell's argument for $z = 1$ [14] is extended to general z. Consider the (maximal) chains in the lattice where nodes are subsets of $\{1, ..., t\}$ and the comparison is "containment". Then each chain contains at most one column of M (for otherwise we have one column contained in another). Two k-subsets C, C' are called s-neighbors if $|C \backslash C'| = |C' \backslash C| = s$.

A chain is said to be blocked by a column if it contains an s-neighbor of the column for some $s \leq \frac{z-1}{2}$. First consider z odd. Consider two columns C and C'. Suppose a chain contains both an s-neighbor of C and an s'-neighbor of C'. Without loss of generality, assume $s \leq s'$. Then $|C \backslash C'| \leq s + s'$. It follows from Corollary 3.7.7 that either $s > (z-1)/2$ or $s' > (z-1)/2$. Therefore a chain cannot be blocked by more than one column of M. So we can associate with C the cost $h(C)$ which is the proportion of chains blocked by C. Consider a node represented by a k-subset. Then it has

$$K_{z-1,k} = \sum_{s=0}^{(z-1)/2} \binom{k}{s}\binom{t-k}{s}$$

s-neighbors for some $s \leq \frac{z-1}{2}$. Hence

$$h(C) = \frac{K_{z-1,k}}{\binom{t}{k}}.$$

It is readily verified that $h(C)$ is minimized when $k = \lfloor t/2 \rfloor$ or $\lceil t/2 \rceil$. In either case, the value of $K_{z-1,k}$ is K_{z-1}. Since the total proportion of chains blocked is at most 1, $1/h(C)$ is an upper bound of n.

Next consider z even. The only difference is that if $|C \backslash C'| = z$ there might exist a chain which contains both a $z/2$-neighbor of C and a $z/2$-neighbor of C'; hence such a chain is multiply blocked.

Let $\mathcal{C}$ denote the set of columns of M each of which has a $z/2$-neighbor in the chain $\mathcal{Z} = \{\emptyset, \{1\}, \{1,2\}, ..., \{1,2,...,t\}\}$. Let $C, C' \in \mathcal{C}$ with $|C| = k$, and $|C'| = k' \geq k$. Then

$$|\{1,2,...,k\} \backslash C| = \frac{z}{2} = |C \cap \{k+1, k+2, ..., t\}|.$$

Let y denote the number of elements in $\{1,2,...,k\} \backslash C$ not in C'. Then C' can leave out at most $z/2 - y$ elements in $C \cap \{1,...,k\}$.

Therefore $|C \backslash C' \cap \{1,...,k\}| \leq z/2 - y$. Since C has only $z/2$ elements in $\{k+1,...,t\}$, $|C \backslash C' \cap \{k+1,...,t\}| \leq z/2$. It follows $|C \backslash C'| \leq z/2 - y + z/2 \leq z$. By Corollary 3.7.7, $|C \backslash C'| \geq z$ which can be realized only if $|C \backslash C' \cap \{1,...,k\}| = |C \backslash C' \cap \{k+1,...,t\}| = z/2$. Namely, C' contains $\{1,...,k\} \backslash C$, is disjoint from $C \cap \{k+1,...,t\}$, and leaves out exactly $z/2$ elements in $C \cap \{1,...,k\}$.

Since C' is a $z/2$-neighbor of $\mathcal{Z}$ which differs from the set $\{1,...,k'\}$ in $z/2$ elements among its first k elements, either $k' = k$, or $k' > k$ but $\{k+1,...,k'\} \subseteq C'$. Note that only one C' of the second type can exist in $\mathcal{C}$. To see this, let C' be a $z/2$-neighbor of $\{1,...,k\}$ and C'' a $z/2$-neighbor of $\{1,...,k''\}$, $k'' \geq k' > k$. Then $|C' \backslash C''| = |C' \backslash C'' \cap \{C \cap \{1,...,k\}\}| \leq \frac{z}{2} < z$.

For $C', C'' \in \mathcal{C}$, let $k'' \geq k'$. Then $k' = k$.

$$\begin{aligned} |C' \backslash C''| &= |C' \backslash C'' \cap \{1,...,k\}| + |C' \backslash C'' \cap \{k+1,...,t\}| \\ &\leq z/2 + z/2 = z, \end{aligned}$$

where equality assumes only if the left-out parts from $C \cap \{1,...,k\}$ are disjoint, and the parts taken from $\{k+1,...,t\}$ are also disjoint. By treating $C \backslash \{1,...,k\}$ as the left-out part of C from $\{1,...,k\}$, then all columns in $\mathcal{C}$ must have disjoint left-out parts from $\{1,...,k\}$ and disjoint parts from $\{k+1,...,t\}$. Therefore the number of $z/2$-neighbors of a given chain is at most $\min\{\frac{k}{z/2}, \frac{t-k}{z/2}\}$ which is maximized at $k = \lfloor t/2 \rfloor$ with the value T.

Associated with C a cost

$$h'(C) = K'_{z-1,k}/\binom{t}{k},$$

where

$$K'_{z-1,k} = K_{z-2,k} + \frac{\text{the number of } z/2\text{-neighbors of } C}{T}.$$

The cost is minimized at $k = \lfloor t/2 \rfloor$ or $\lceil t/2 \rceil$. In either case $K'_{z-1,k} = K_e$ and $1/h'(C)$ is an upper bound of n. □

Corollary 3.7.10 *If* $S(\lfloor t/2 \rfloor - 1, \lfloor t/2 \rfloor, t)$ *exists, then it yields an optimal* $(1;2)$*-disjunct matrix.*

Proof. The number of elements in $S(\lfloor t/2 \rfloor - 1, \lfloor t/2 \rfloor, t)$ is

$$\begin{aligned}\binom{t}{\lfloor t/2 \rfloor - 1}\Big/\binom{\lfloor t/2 \rfloor}{\lfloor t/2 \rfloor - 1} &= \binom{t}{\lfloor t/2 \rfloor}\Big/(t - \lfloor t/2 \rfloor + 1) \\ &= \binom{t}{\lfloor t/2 \rfloor}\Big/(1 + \lceil t/2 \rceil) = \binom{t}{\lfloor t/2 \rfloor}\Big/K_1.\end{aligned}$$

□

Let $\mathcal{F}$ be a family of subsets of $\{1, ..., t\}$. $\mathcal{F}$ is called a $(2, z-1)$-cover of a k-subset B if

1. $b \in \mathcal{F}$ and $b \subset b' \subseteq B$ implies $b' \in \mathcal{F}$.

2. For every $b \subseteq B$ with $|B \backslash b| \leq z - 1$, at least one part of every 2-partition of b is in $\mathcal{F}$. Note that $b \subseteq B$ and $|b| \geq k - z + 1$ implies $b \in \mathcal{F}$ since b contains both parts of a partition.

Lemma 3.7.11 *M is $(2;z)$-disjunct if and only if for each column C the private (with respect to M) subsets of C form a $(2, z-1)$-cover of C.*

Proof. Let $b \subset C$ with $|C \backslash b| \leq z - 1$. From Corollary 3.7.7, b is private for if $b \subset C'$, then $|C \backslash C'| \leq z - 1$. If there exists a partition of b into two nonprivate parts, then the union s of the two columns containing them will lead to $|C \backslash U(s)| \leq z - 1$, violating Lemma 3.7.6. This proves the necessity.

To prove sufficiency, assume that s is a set of two columns such that $|C \backslash U(s)| \leq q$. Set $b = U(s) \cap C$. Since $b \in \mathcal{F}$ by condition 1, condition 2 of $(2, z-1)$-cover is violated. □

For $\mathcal{F}$ a $(2, z - 1)$-cover of C, define $h(C, \mathcal{F})$ to be the proportion of all chains which intersect $\mathcal{F}$.

Theorem 3.7.12 *For any k-subset B, $z \le k < t-1$, and $\mathcal{F}$ a $(2, z-1)$-cover of B,*

$$h(B, \mathcal{F}) \ge \binom{2x+z-2}{x} / \binom{t}{x}$$

where $x = \lfloor (k-z+2)/2 \rfloor$. Equality is achieved if and only if $k-z$ is even and $\mathcal{F} = \mathcal{F}^* = \{b \subseteq B : |b| \ge x\}$.

Proof. Suppose that $k-z$ is even, *i.e.*, $k = 2x+z-2$. Then every 2-partition of any $(k-z+1)$-subset of B contains a part b with $|b| \ge x$. Hence $\mathcal{F}^*$ is a $(2, z-1)$-cover of B. It is easily verified that

$$h(B, \mathcal{F}^*) = \binom{2x+z-2}{x} / \binom{t}{x}.$$

Suppose that $\mathcal{F}$ achieves minimum in $h(B, \mathcal{F})$. Let s denote the largest integer such that there exists some $b \subset B$ with $|b| = x+s$ and $b \notin \mathcal{F}$. If $s \ge x-1$, $|B \backslash b| \le z-1$ and $b \in \mathcal{F}$ followed from the definition of $(2, z-1)$-cover. Hence $s < x-1$. If $s < 0$, then $|b| \le x-1$, and $\mathcal{F}^* \subseteq \mathcal{F}$. It follows $h(B, \mathcal{F}^*) \le h(B, \mathcal{F})$. Therefore we may assume $s \ge 0$. Suppose there exists some $b \in \mathcal{F}$ such that $|b| < x-s-1$. Let $b \subseteq b' \subseteq B$ such that $|B \backslash b'| \le z-1$. Then $|b' \backslash b| > k-z+1-(x-s-1) = x+s$. Hence $b' \backslash b \in \mathcal{F}$ and $\mathcal{F} \backslash \{b\}$ is still a $(2, z-1)$-cover. Therefore we may assume $\mathcal{F}$ contains no such b.

Let f_r and $\bar{f}_r$ denote the number of r-subsets of B in and not in $\mathcal{F}$, respectively. Then each such r-subset appears in $\binom{k-r}{z-1}$ $(k-z+1)$-subsets b of B since the $z-1$ indices of B not in b must not be chosen from the r-subset to keep it intact in b. Clearly, the number of $(x+s)$-subsets of b not in $\mathcal{F}$ cannot exceed the number of $(x-s-1)$-subsets of b in $\mathcal{F}$ due to the $(2, z-1)$-cover property. Hence

$$\binom{k-x-s}{z-1} \bar{f}_{x+s} \le \binom{k-x+s+1}{z-1} f_{x-s-1},$$

or

$$\frac{\bar{f}_{x+s}(k-x-s)!(x+s)!}{f_{x-s-1}(k-x+s+1)!(x-s-1)!} \le 1.$$

Construct $\mathcal{F}'$ from $\mathcal{F}$ by removing all $(x-s-1)$-sets and adding any missing $(x+s)$-subsets of B. Then $\mathcal{F}'$ is also a $(2, z-1)$-cover of B. We compute:

$$\frac{h(B, \mathcal{F}')}{h(B, \mathcal{F})} = \frac{\text{the portion of chains which contain a } (x+s)\text{-subset of } B \text{ but no element of } \mathcal{F}}{\text{the portion of chains which contain a } (x-s-1)\text{-subset of } B \text{ but no other element of } \mathcal{F}}. \tag{3.7.1}$$

Let b be a $(x+s)$-subset not in $\mathcal{F}$. Then its lower paths (from this node to node $\emptyset$) does not intersect $\mathcal{F}$. An upper path does not intersect $\mathcal{F}$ if the parent node of b

is not contained in B. If it is contained in B, then it intersects $\mathcal{F}$ by our assumption that all subsets of B with size $> x+s$ are in $\mathcal{F}$. A parent node of b is simply $b \cup \{y\}$, for some $y \notin b$. So $b \cup \{y\} \not\subseteq B$ if and only if $y \notin B$, and there are $t-k$ choices of it. So the proportion of chains going through b but not in $\mathcal{F}$ is $(t-k)/(t-x-s)$. Since this is true for every $(x+s)$-subset b not in $\mathcal{F}$, the first term in (3.7.1) equals

$$\frac{t-k}{t-x-s}\bar{f}_{x+s}/\binom{t}{x+s}.$$

On the other hand, let b now denote a $(x-s-1)$-subset in $\mathcal{F}$. Then an upper path does not intersect $\mathcal{F}$ if and only if the parent node of b is not contained in B, and the proportion is exactly $(t-k)/(t-x+s-1)$. A lower path does not intersect $\mathcal{F}$ since we assume $\mathcal{F}$ contains no subset of size $< x-s-1$. Therefore the proportion of chains going through b but not containing any other element of $\mathcal{F}$ is $(t-k)/(t-x+s+1)$. Consequently, the second term in (3.7.1) equals

$$\frac{t-k}{t-x+s+1}f_{x-s-1}/\binom{t}{t-x+s+1}.$$

Thus

$$\begin{aligned}\frac{h(B,\mathcal{F}')}{h(B,\mathcal{F})} &= \frac{(t-k)\bar{f}_{x+s}}{(t-x-s)\binom{t}{+s}}/\frac{(t-k)f_{x-s-1}}{(t-x+s+1)\binom{t}{t-x+s+1}} \\ &= \frac{\bar{f}_{x+s}(t-x+s-1)!(x+s)!}{f_{x-s-1}(t-x+s)!(x-s-1)!} \\ &< \frac{\bar{f}_{x+s}(k-x-s)!(x+s)!}{f_{x-s-1}(k-x+s+1)!(x-s-1)!} \le 1,\end{aligned}$$

contradicting the minimality of $\mathcal{F}$. Therefore $s<0$ and $\mathcal{F}^*$ achieves minimum.

Suppose $k = 2x+z-1$, $z>1$. Then for any $y \in B$, $\{b \in \mathcal{F} : y \notin b\}$ is a $(2, z-2)$-cover for $B\backslash\{y\}$. So the lemma follows from the case $k = 2x+z-2$. If $z=1$, then every partition of B has one part in $\mathcal{F}$. Consider the partition of B into two x-sets. Then at least $\binom{2x}{x}/2 = \binom{2x-1}{x}$ x-subsets are in $\mathcal{F}$. Again, we can show that $\mathcal{F}$ contains all $(x+1)$-subsets but no $(x-1)$-subsets. Thus

$$h(B,\mathcal{F}) \ge \binom{2x-1}{x}/\binom{t}{x} \text{ and } n \le \binom{t}{x}/\binom{2x-1}{x}$$

(also proved in Lemma 2.4.3). □

Theorem 3.7.13 *A $t \times n$ $(2;z)$-disjunct matrix satisfies*

$$n \le \binom{t}{x}/\binom{2x+z-2}{x},$$

where x is the smallest integer satisfying

$$t \le 5x+2+(z-1)(z-2)/(x+z-1)(x = \lceil (t-2)/5 \rceil \textit{ for } z=1,2). \qquad (3.7.2)$$

Proof. Define $F(x) = \binom{2x+z-2}{x} / \binom{t}{x}$. Then

$$\frac{F(x+1)}{F(x)} = \frac{(2x+z)(2x+z-1)}{(t-x)(x+z-1)},$$

which exceeds one if and only if (3.7.2) is not satisfied. Hence $F(x)$ is minimized by (3.7.2) and n is upper bounded by $\frac{1}{F(x)}$. □

References

[1] D. J. Balding and D.C. Torney, Optimal pool designs with error detection, *J. Combin. Thy., Series A*, 74 (1996) 131-140.

[2] E. Barillot, B. Lacroix and D. Cohen, Theoretical analysis of library screening using a n-dimensional pooling strategy, *Nucl. Acids Res.*, (1991), no. 19, 6241-6247.

[3] B. S. Chlebus and D. R. Kowalski, Almost optimal explicit selector, *Lecture Notes in Computer Science* 3623 (Springer-Verlag, 2005) 270-280.

[4] C. Colbourn, D. Hoffman and R. Reed, A new class of group divisible designs with block size three, *J. Combin. Thy.*, 59 (1992) 73-89.

[5] D.-Z. Du, F. K. Hwang, W. Wu and T. Znati, A new construction of transversal designs, *Journal of Computational Biology*, to appear.

[6] D.-Z. Du, F. K. Hwang, W. Wu and T. Znati, An improvement on construction of transversal designs, manuscript.

[7] A. J. Hoffman and R. R. Singleton, On Moore graphs with diameters 2 and 3, *IBM J. Res. Develop.*, 4 (1960) 497-504.

[8] W.T. Huang and F. K. Hwang, private communication.

[9] F. K. Hwang, An isomorphic factorization of the complete graph, *Journal of Graph Theory*, 19 (1995) no. 3, 333-337.

[10] F. K. Hwang and V. T Sós, Non-adaptive hypergeometric group testing, *Studia Scient. Math. Hungarica*, 22 (1987) 257-263.

[11] P. Indyk, Explicit constructions of selecors and related combinatorial structures with applications, *Proceedings of 13th ACM-SIAM Symposium on Discrete Algorithms (SODA)*, 2002, pp. 697-704.

[12] W. H. Kautz and R.C. Singleton, Nonrandom binary superimposed codes, *IEEE Trans. Inform. Thy.*, 10 (1964) 363-377.

[13] B. Lindstrom, Determination of two vectors from the sum, *J. Combin. Thy.*, A6 (1969) 402-407.

[14] D. Lubell, A short proof of Sperner's lemmas, *J. Combin. Theory*, 1 (1966) 299.

[15] Q. A. Nguyen and T. Zeisel, Bounds on constant weight binary superimposed codes, *Probl. Control & Inform. Thy.*, 17 (1988) 223-230.

[16] D. Raghavarao, *Constructions and Combinatorial Problems in Designs of Experiments*, (Wiley, New York, 1971).

[17] V. Rödl, On a packing and covering problem, *Euro., J. Combi.*, 5 (1984).

[18] S. Roman, *Coding and Information Theory*, (Springer-Verlag, New York, 1992).

[19] G. M. Saha, H. Pesotan and B. L. Raktoc, Some results on t-complete designs, *Ars Combin.*, 13 (1982) 195-201.

[20] J. Schönheim, On maximal systems of k-tuples, *Studia Sci. Math. Hungar*, 1 (1966) 363-368.

[21] D. J. Schultz, M. Parnes and R. Srinivasan, Further applications of d-complete designs to group testing, *J. Combin. Inform. & System Sci.*, 8 (1993) 31-41.

[22] R. R. Singleton, Maximum distance q-nary codes, *IEEE Trans. Inform. Thy.*, 10 (1964) 116-118.

[23] A. Ta-Shma, C. Uman and D. Zuckerman, Loss-less condensers, unbalanced expanders, and extractors, *Proceedings of 33rd ACM Symposium on Theory of Computing (STOC)*, 2001, pp. 143-152.

[24] F. Vakil and M. Parnes, On the structure of a class of sets useful in non-adaptive group-testing, *J. Statist. Plan. & Infer.*, 39 (1994) 57-69.

[25] F. Vakil, M. Parnes and D. Raghavarao, Group testing with at most two defectives when every item is included in exactly two group tests, *Utilitas Math.*, 38 (1990) 161-164.

[26] C. A. Weideman and D. Raghavarao, Some optimum non-adaptive hypergeometric group testing designs for identifying defectives, *J. Statist. Plan. Infer.*, 16 (1987) 55-61.

[27] C. A. Weideman and D. Raghavarao, Nonadaptive hypergeometric group testing designs for identifying at most two defectives, *Commun. Statist.*, 16A (1987), 2991-3006.

[28] V. Zinoviev, Cascade equal-weight codes and maximal packing, *Probl. Control & Inform. Thy.*, 12 (1983) 3-10.

4

Deterministic Designs from Partial Orders

In this chapter, we introduce a class of designs for the d-disjunct matrix with geometric and algebraic tools, including simplicial complex, linear algebra on finite field, and lattice.

4.1 Subset Containment Designs

Macula [7, 9] proposed an interesting construction for d-disjunct matrices. Let $SCD(m, d, k)$ be a binary matrix with rows indexed by all d-subsets of $\{1, 2, ..., m\}$ and columns indexed by all k-subsets of $\{1, 2, ..., m\}$ where $d \leq k$. $SCD(m, d, k)$ has a 1-entry at cell (i, j) if and only if the i d-subset is contained by the jth k-subset. This design is called a *subset containment design (SCD)* with parameters (m, k, d).

Consider a column C_{S_0} and d other distinct columns $C_{S_1}, \ldots, C_{S_d}$. Since S_0, S_1, $\ldots$, S_d are distinct k-subsets, $S_0 \setminus S_1$, $\ldots$, $S_0 \setminus S_d$ are all nonempty. One can choose a_1 from $S_0 \setminus S_1$, $\ldots$, a_d from $S_0 \setminus S_d$ and put them together with possibly some other elements of S_0 (in case that $a_1, \ldots, a_d$ are not distinct) to form a d-subset I of S_0. At the row with index I, column C_{S_0} has an 1-entry and all columns $C_{S_1}, \ldots, C_{S_d}$ have 0-entries. Therefore, C_{S_0} cannot be contained in the union of $C_{S_1}, \ldots, C_{S_d}$. This argument shows that the following theorem holds.

Theorem 4.1.1 *$SCD(m, d, k)$ is d-disjunct.*

Macula also defined $SCD^*(m, d, k)$ by adding n rows $\bar{1}$, $\bar{2}$, ..., $\bar{n}$, where $\bar{i}$ has 0 in each column whose label contains i and 1 elsewhere. He [10] proved that $SCD^*(m, d, k)$ has some error-correcting ability. Here we report a modification observed in Hwang [5].

Theorem 4.1.2 *Suppose $k - d \geq 3$. Then $SCD^*(m, d, k)$ is $(\bar{d}, 4)$-separable.*

Proof. Let $s = (C_{j_1}, ..., C_{j_s})$ and $s' = (C_{h_1}, ..., C_{h_r})$ be two distinct sample points, with $r \leq s \leq d$. Without loss of generality, assume $C_{j_1} \notin s'$. We consider three cases:

Case 1. $r \leq d-1$. Then for each $1 \leq i \leq r$, there exists an $x_i \in C_{j_1} \backslash C_{h_i}$. So the set $\{x_1, ..., x_r\}$ is in C_{j_1}, but not in C_{h_i}, $1 \leq i \leq r$; or equivalently, any row whose label containing $\{x_1, ..., x_r\}$ intersects C_{j_1}, but not $C_{h_i}, 1 \leq i \leq r$. Since $r \leq d-1$ and $k-d \geq 3$, there exist at least four such rows.

Case 2. $r = d$ and $|s \cap s'| = d-1$. Without loss of generality, assume C_{j_1} is the unique column not in s'. The set $X = \{x_1, x_2, ..., x_r = x_d\}$ still yields a row which intersects C_{j_1} but not C_{h_i}, $1 \leq i \leq d$. If $x_1 \in C_{j_1} \backslash C_{h_i}$ for some $2 \leq i \leq d$, then we can set $x_i = x_1$, and reduce $|X|$ to $d-1$. The argument of case 1 also works here. We assume that $x_1 \in C_{h_i}$ for all $2 \leq i \leq d$. Since $C_{j_i} = C_{h_i}$ for $2 \leq i \leq d, \bar{x}_1$ does not appear in any column of s, but appear in C_{h_1} of s'. Namely, the row labeled by $\bar{x}_1$ intersects s' but not s. Thus we obtain two rows which distinguish s from s'. The other two rows are obtained by symmetry, *i.e.*, using C_{h_1} instead of C_{j_1}.

Case 3. $r = d$ and $|s \cap s'| < d-1$. There are two columns C_{j_1} and C_{j_2} not in s', which generates two d-sets $X = \{x_1, ..., x_d\}$ and $Y = \{y_1, ..., y_d\}$ which intersect s but not s'. Let $z \in C_{j_1} \backslash C_{j_2}$. If $z \notin C_{h_i}$ for some $1 \leq i \leq d$, we set $x_i = z$, hence $X \neq Y$. If $z \in C_{h_i}$ for all $1 \leq i \leq d$ then z is in every column of s', but not in C_{j_2}. Hence the row labeled by $\bar{z}$ intersects s but not s'. Therefore in either case we obtain two rows which distinguish s from s'. The other two rows are obtained by the symmetry of s and s'. □

Hwang [5] also showed that both the condition $k - d \geq 3$ and the Hamming distance 4 cannot be improved.

An immediate corollary is that

Corollary 4.1.3 *$SCD^*(m, d, k)$ is an 1-error-correcting d-disjunct matrix.*

Fu and Hwang [4] showed that if the column indices are chosen judiciously, instead of arbitrarily, then d can be dramatically increases.

Let $\mathcal{P}$ be a Steiner (t, v, k)-packing where $|\mathcal{P}| \geq n$. Consider $SCH(m, d, k)$ whose n columns have indices chosen from subsets of $\mathcal{P}$. Since each column has weight $\binom{k}{d}$ and each pair of columns intersect at most $\binom{t-1}{d}$. By Lemma 2.3.11, we have

Theorem 4.1.4 *Suppose $\binom{k}{d} = d\binom{t-1}{d} + q$ for some $1 \leq q < \binom{t-1}{d}$. Then $SCH(m, d, k)$ is $(d; q)$-disjunct.*

Note that for $d = 1$, Theorem 4.1.4 is reduced to Theorem 3.1.2.

4.2 Partial Order of Faces in a Simplicial Complex

Macula's SCD construction is generalized to simplicial complexes by Park, Wu, Wu, Liu and Zhao [14].

The simplex is a generalization of the triangle in different dimensions. The triangle is called a two-dimensional simplex and the tetrahedron is called a three-dimensional simplex. A zero-dimensional simplex is a point and a one-dimensional simplex is a line-segment. In general, an n-dimensional *simplex* is a polytope with $n+1$ vertices in n-dimensional space, which is not degenerated, i.e., not contained in any n-dimensional hyperplane.

The simplex has an important property that any subset of its vertices is the vertex set of a simplex, which is called a *face* of the original simplex. A family of simplexes is called a *geometric simplicial complex* if any face of a simplex in the family is also in the family. This means that the intersection of two different simplexes is a face for both simplexes.

Consider a finite set V. A family Δ of subsets of V is called an (abstract) *simplicial complex* if any subset of a member in Δ is also a member. Each member of Δ is called a *face* of Δ. The *dimension* of a face A is $|A|-1$. Any face of dimension 0 is called a *vertex*. For example, consider a set $V = \{a, b, c, d\}$. The following family is a simplicial complex on V:

$$\begin{aligned}\Delta &= \{\emptyset, \{a\}, \{b\}, \{c\}, \{d\},\\ &\quad \{a,b\}, \{b,c\}, \{c,d\}, \{d,a\}, \{a,c\},\\ &\quad \{a,b,c\}, \{a,c,d\}\}.\end{aligned}$$

$\{a,b,c\}$ is a face of dimension two and the empty set $\emptyset$ is a face of dimension -1.

Associate each vertex x_i with the standard basis vector e_i in the Euclidean space R^n. Then the *geometric representation* $\hat{\Delta}$ of Δ is the union of all the convex hulls $\hat{A} = conv(\{e_i \mid x_i \in A\})$ for $A \in \Delta$. The geometric representation of the simplicial complex in the above example is shown in Fig. 4.1. It is often helpful to think of a

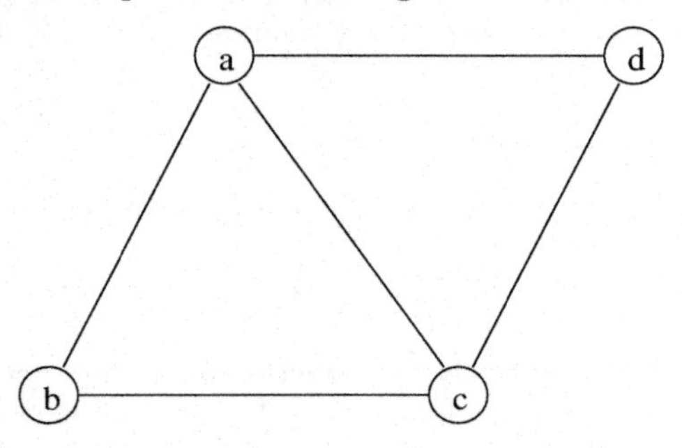

Figure 4.1: The geometric representation of a simplicial complex.

simplicial complex in term of its geometric representation. For example, the definition of the dimension of a face is very meaningful in the geometric representation.

Assume $k \geq d \geq 1$. Let Δ be a simplicial complex with d-faces $f_1, \ldots, f_m$ and k-faces $g_1, \ldots, g_n$. Define matrix $M(\Delta, d, k) = (a_{ij})$ by

$$a_{ij} = \begin{cases} 1 & \text{if } f_i \subset g_j \\ 0 & \text{otherwise} \end{cases}$$

Theorem 4.2.1 *Let $2 \leq d \leq k$. Then $M(\Delta, d, k)$ is a d-disjunct matrix.*

Proof. Consider any set of $d+1$ distinct columns $C_0, C_1, \ldots, C_d$ corresponding to, respectively, $d+1$ distinct k-faces $A_0, A_1, \ldots, A_d$. Choose $a_1 \in A_0 \setminus A_1, \ldots, a_d \in A_0 \setminus A_d$. Set $I = \{a_1, \ldots, a_d\}$. If $|I| = d$, then I is a d-face in A_0, but not in $A_1, \ldots, A_d$. If $|I| < d$, then choose $d - |I|$ elements from $A_0 - I$, together with elements in I to form a d-face $I' \supset I$. Clearly, I' is in A_0, but not in $A_1, \ldots, A_d$. This means that in either case, we can find a d-face such that at the corresponding row, column C_0 has an 1-entry and all columns $C_1, \ldots, C_d$ have 0-entries. Therefore, C_0 cannot be contained in the union of $C_1, \ldots, C_d$. Hence, $M(\Delta, d, k)$ is d-disjunct. □.

Ngo and Du [11] found that $M(\Delta, d, k)$ has some error-correcting ability under certain conditions. Their result is based on the following lemma.

Lemma 4.2.2 *Suppose a graph G has d edges and more than d vertices. Then G has at least $d+1$ vertex-covers of size d. (Recall that a vertex-cover is a subset of vertices such that every edge is incident to a vertex in the subset.)*

Proof. Let U be the set of vertices each with degree at least one. If $|U| \leq d+1$, then add $d+1-|U|$ isolated vertices into U to form a set U' of $d+1$ vertices. For any $v \in U'$, $U' - \{v\}$ is a vertex cover of size d. Therefore, $d+1$ vertex covers of size d is found.

Now, consider $|U| \geq d+2$. In this case, all d edges must be distributed into c $(\geq |U| - d = k)$ connected components $C_1, \ldots, C_c$. Note that for any $1 \leq i_1 < \cdots < i_k \leq c$, there are $|C_{i_1}| \cdots |C_{i_k}|$ vertex covers of size d in the form $U - \{v_{i_1}, \ldots, v_{i_k}\}$ for $v_{i_1} \in V(C_{i_1}), \ldots, v_{i_k} \in V(C_{i_k})$. Therefore, the number of vertex covers of size d for G is at least

$$\sum_{1 \leq i_1 < \cdots < i_k \leq c} |C_{i_1}| \cdots |C_{i_k}| \geq \sum_{i=1}^{c} |C_i| \geq d+2.$$

Here, we use an inequality as follows: Suppose $x_i \geq 2$ for $i = 1, \ldots, c$. Then

$$\sum_{1 \leq i_1 < \cdots < i_k \leq c} x_{i_1} \cdots x_{i_k} \geq \sum_{i=1}^{c} x_i.$$

This inequality can be proved easily by induction on c. For $c = 1$, it is trivial. For $k = 1$, it is also trivial. Now, consider $c \geq k \geq 2$. Then, we have

$$\begin{aligned} \sum_{1 \leq i_1 < \cdots < i_k \leq c} x_{i_1} \cdots x_{i_k} &\geq x_c \sum_{1 \leq i_1 < \cdots < i_{k-1} \leq c-1} x_{i_1} \cdots x_{i_k} \\ &\geq x_c \sum_{i=1}^{c-1} x_i \geq \sum_{i=1}^{c} x_i. \end{aligned}$$

□

Park *et al.* also obtained some results on error-correcting by extending the above lemma. Chang *et al.* [2] improved the extension through the lemma.

Lemma 4.2.3 *For $2 \leq d < n \leq 2^d$, a graph G with d edges and n vertices has at least n vertex-covers of size d.*

Proof. Let U be the set of vertices with degree at least one.

Case 1. $|U| \leq d$. For any $v \in U$, $U - \{v\}$ is a vertex-cover of size $|U| - 1$. Take any $d - |U| + 1$ isolated vertices to form a vertex-cover of size d. The number of choices is

$$|U| \cdot \binom{n - |U|}{d - |U| + 1} \geq |U|(n - |U|) \geq n - 1.$$

Moreover, U together with $d - |U|$ isolated vertices forms one additional vertex-cover of size d.

Case 2. $|U| = 2d$. Then all d edges are disjoint. There are totally $2^d \geq n$ vertex-covers of size d.

Case 3. $d + 1 \leq |U| \leq 2d - 1$. Let $C_1, ..., C_c$ be nonempty components of G. For each C_i, denote by d_i the number of edges and by n_i the number of vertices.

Subcase 3.1. $n_i = d_i + 1$ for all $i = 1, ..., c$. In this subcase, every C_i is a tree. Deleting a vertex from each C_i yields a vertex-cover of size $d_1 + \cdots + d_c = d$. In this way, we obtain

$$\prod_{i=1}^{c} n_i \geq \sum_{i=1}^{c} i = |U|$$

vertex-covers of size d. Since $|U| \leq 2d - 1$, there must exist a component C_i having more than one edges. In this component, there must exist a vertex v with degree at least two. Suppose u and w are adjacent to v. Deleting w and u from this component and a vertex from each of other components would result in a vertex-cover of size $d - 1$. Now, adding an isolated vertex in, we obtain $n - |U|$ more vertex-covers of size d.

Subcase 3.2. For some i, $n_i \leq d_i$. In this subcase, $|U| - d + 1 \leq c$. Choose $k = |U| - d + 1$ components and delete a vertex from each of them. The number of vertex-covers of size $d - 1$ obtained in this way is

$$\begin{aligned} & \sum_{1 \leq i_1 < \cdots < i_k \leq c} n_{i_1} \cdots n_{i_k} \\ \geq & \sum_{1 \leq i_1 < \cdots < i_k \leq c} (n_{i_1} + \cdots + n_{i_k}) \\ \geq & \sum i = 1^c n_i = |U|. \end{aligned}$$

Similarly, we can also obtain $|U|$ vertex-covers of size d from U. Now, if $n = |U|$, then it is done. If $n > |U|$, then $|U|$ vertex-covers of size $d - 1$ combined with $n - |U|$ isolated vertices yield additional $|U|(n - |U|)$ vertex-covers of size d. Therefore, totally, we have

$$|U| + |U|(n - |U|) \geq n$$

vertex-covers of size d. □

The following theorem is established from the above two lemmas.

Theorem 4.2.4 *Let Δ be a simplicial complex satisfying condition that for any two distinct k-faces A and A', $|A \setminus A'| \geq 2$. Then the following holds:*
(a) If $1 \leq d < k$, then $M(\Delta, d, k)$ is $(d; d+1)$-disjunct.
(b) If $2 \leq d < k \leq 2^d$, then $M(\Delta, d, k)$ is $(d; k)$-disjunct.

Proof. (a) For any column C_0 and other d distinct columns $C_1, \ldots, C_d$, we show that there exist $d+1$ rows such that at each of these $d+1$ rows, C_0 has 1-entry and all $C_1, \ldots, C_d$ have 0-entry. To this end, one needs to show that for any $d+1$ distinct k-faces $A_0, A_1, \ldots, A_d$, there exist $d+1$ distinct d-faces in A_0, but not in $A_1, \ldots, A_d$. To do so, one chooses two distinct elements u_i and v_i from each $A_0 \setminus A_i$ for $i = 1, \ldots, d$. Construct a graph G with vertex set A_0 and d edges (u_i, v_i) for $i = 1, \ldots, d$. If only $d' < d$ edges are distinct, we add $d - d'$ arbitrary distinct edges. By Lemma 4.2.2, G has at least $d+1$ vertex covers of size d. Each of those $d+1$ vertex-covers is actually a d-face in A_0, but not in $A_1, \ldots, A_d$.

For (b), the argument is similar to the proof of (a) except that Lemma 4.2.3 is used to replace Lemma 4.2.2. □

Macula [10] presented an interesting family of finite sets. Consider set $N = \{1, 2, ..., n\}$ where $n = rq$. Denote

$$D(s_1, s_2, ..., s_r) = \{s_1, q + s_2, 2q + s_3, ..., (r-1)q + s_r\}$$

for $s_1, s_2, ..., s_r \in Z_q = \{0, 1, ..., q-1\}$. Let

$$B(q, r) = \{D(s_1, s_2, ..., s_r) \mid s_1 + s_2 + \cdots + s_r \equiv 0 \pmod q\}.$$

Then $|B(q, r)| = q^{r-1}$. Note that the family $B(q, r)$ has a property that for any distinct $D(s_1, s_2, ..., s_r), D(s'_1, s'_2, ..., s'_r) \in B(q, r)$, $|D(s_1, s_2, ..., s_r) \setminus D(s'_1, s'_2, ..., s'_r)| \geq 2$. In fact, vectors $(s_1, s_2, ..., s_r)$ and $(s'_1, s'_2, ..., s'_r)$ are either identical or have at least two different components.

Let $\Delta(q, r) = \{A \mid A \subseteq D \in B(q, r)\}$. Then $\Delta(q, r)$ is a simpicial complex and $B(q, r)$ form a family of r-faces satisfying the condition in Theorem 4.2.4.

Following Macula's $SCH^*(m, d, k)$ construction, we define matrix $M^*(\Delta, d, k)$ by adding $|X|$ rows to $M(\Delta, d, k)$ as follows: These $|X|$ rows are labeled by $\bar{x}$ for all $x \in X$ and the entry at cell $(\bar{x}, B_j)$ is 1 if and only if $x \notin B_j$. For simplicity, we will call rows with labels $A_1, ..., A_t$ in *the first part* and rows with label $\bar{x}$ in *the secod part.*

Wu *et al.* [15] found that both $M(\Delta, d, k)$ and $M^*(\Delta, d, k)$ are very good for error-tolerance when applied to samples with at most $d-1$ positive clones.

Theorem 4.2.5 *For $k > d \geq 2$, $M(\Delta, d, k)$ is $(d-1, k-d)$-disjunct.*

Proof. Consider d distinct columns labeled by d distinct k-faces $B_0, B_1, ..., B_{d-1}$. Choose $a_1 \in B_0 \setminus B_1, ..., a_{d-1} \in B_0 \setminus B_{d-1}$. Set $I = \{a_1, ..., a_{d-1}\}$. Now, for any subset I' of $d - |I|$ vertices in $B_0 - I$, $I \cup I'$ is a d-face such that on the row with label $I \cup I'$, column B_0 has a 1-entry and columns $B_1, ..., B_{d-1}$ have 0-entry. Therefore, column B_0 contains at least $\binom{k-|I|}{d-|I|}$ 1-entries not appearing in the union of columns $B_1, ..., B_{d-1}$. Finally, since $|I| \le d-1$, we have $\binom{k-|I|}{d-|I|} \ge k-d+1$. □

The next result shows that the $M^*(\Delta, d, k)$ can have more error-correcting ability.

Theorem 4.2.6 *For $k > d \ge 1$,*

$$H_{d-1}(M^*(\Delta, d, k)) \ge 2\min(\binom{k-d+2}{2}, 2(k-d)+1, k)$$

and

$$H_{\overline{d-1}}(M^*(\Delta, d, k)) \ge \min(\binom{k-d+2}{2}, 4(k-d)+2, 2k).$$

Proof. To show the first inequality, let U be a union of $d-1$ columns with labels $B_1, ..., B_{d-1}$ and U' another union of $d-1$ columns with labels $B'_1, ..., B'_{d-1}$. We claim that either

(a) U contains $\min(\binom{k-d+2}{2}, 2(k-d)+1)$ 1-entries in the first part of rows, not in U',

or

(b) U contains $k-d+1$ 1-entries in the first part of rows and $d-1$ 0-entries in the second part of rows, not in U'.

To prove our claim, choose $a_{i1} \in B_i \backslash B'_1, ..., a_{i,d-1} \in B_1 \backslash B'_{d-1}$. Let $I_i = \{a_{i1}, ..., a_{i,d-1}\}$.

If $|I_i| \le d-2$, then we can find $\binom{k-d+2}{2}$ $(\le \binom{k-|I_i|}{d-|I_i|})$ d-faces in B_i, containing I_i. They label $\binom{k-d+2}{2}$ rows in the first part satisfying (a).

Thus, we may assume $I = I_i$ for all $i = 1, ..., d-1$ and $|I| = d-1$. Then, we can find $k-d+1$ d-faces in B_1 or $\cdots$ or B_{d-1} containing I, which gives $k-d+1$ rows in the first part on which all columns $B'_1, ..., B'_{d-1}$ have 0-entries and at least one of columns $B_1, ..., B_{d-1}$ has 1-entry. For each $a \in I$, on the row with label $\bar{a}$, all columns $B_1, ..., B_{d-1}$ have 0-entries and at least one of columns $B'_1, ..., B'_{d-1}$ has 1-entry. Therefore, (b) holds.

In case (a), U and U' disagree on at least $\min(\binom{k-d+2}{2}, 2(k-d)+1)$ rows and in case (b), U and U' disagree on at least k rows. Note that by exchanging the roles of

U and U', we would obtain at least $\min(\binom{k-d+2}{2}, 2(k-d)+1, k)$ additional rows at which U' and U disagree each other. Therefore,

$$H_{d-1}(M^*(\Delta, d, k)) \geq 2\min(\binom{k-d+2}{2}, 2(k-d)+1, k).$$

To prove the second inequality, we need to consider one more case that either U or U' is a union of at most $d-2$ columns. In this case, we can obtain $\binom{k-d+2}{2}$ rows in the first part of the right-hand side of the above inequality, where U and U' disagree with each other. However, we cannot double the number of rows by exchanging U and U'. Therefore,

$$\begin{aligned} H_{\overline{d-1}}(M^*(\Delta, d, k)) &\geq \min(H_{d-1}(M^*(\Delta, d, k)), \binom{k-d+2}{2}) \\ &= \min(\binom{k-d+2}{2}, 4(k-d)+2, 2k). \end{aligned}$$

□

A d-disjunct matrix without an isolated column (i.e. no pool is singleton) is also $(d-1;2)$-disjunct [3] and in general, this is best possible. For example, there exists a 12×16 2-disjunct matrix without isolated column in (page 147 of [3]) that every column has exactly three 1-entries. This matrix cannot be $(1;3)$-disjunct because for two columns having one 1-entry in common, one cannot have three 1-entries not in the other one.

Theorem 4.2.5 indicates that $M(\Delta, d, k)$ is really good for error-tolerance as long as $k-d$ is large and the number of positives clones does not exceed $d-1$.

For matrix $M^*(\Delta, d, k)$, Theorem 4.2.6 indicates that a similar thing happens. In fact, if we apply $M^*(\Delta, d, k)$ to a sample with up to d instead of $d-1$ positive clones, then the capability for error-tolerance would be quite small. The following theorem states such a result. Indeed, Macula's design [8] is a special case of $M^*(\Delta, d, k)$ and Hwang [5] showed that the inequalities in the theorem are best possible for Macula's design. Hence, they are also best possible to $M^*(\Delta, d, k)$.

Theorem 4.2.7 *(a) For $k > d \geq 1$, $H_d(M^*(\Delta, d, k)) \geq 4$. (b) For $k-d \geq 3$, $H_{\bar{d}}(M^*(\Delta, d, k)) \geq 4$.*

The proof is similar to that of Theorem 4.1.2.

4.3 Monotone Graph Properties

For the simplicial complex, the monotone graph property induces an important family of examples.

Any graph property can be described by the set of all graphs having the property. Note that throughout this section, the terminology "subgraph" only applies to those graphs with the same vertex set as the original graph and a subset of edges. A k-subgraph is a subgraph with k edges.

A graph property is *monotone increasing* if a graph containing a subgraph with the property must have the property. A graph property is *monotone decreasing* if every subgraph of any graph with the property also has the property. A graph property is *monotone* if it is either monotone increasing or monotone decreasing.

Let G be a graph. Define

$$\Delta_{\mathcal{P}}^G = \{E(H) \mid H \text{ is a subgraph of } G, \text{ satisfying } \mathcal{P}\}$$

for a monotone decreasing property $\mathcal{P}$, and

$$\Delta_{\mathcal{P}}^G = \{E(H) \mid H \text{ is a subgraph of } G, \text{ not satisfying } \mathcal{P}\}$$

for a monotone increasing property $\mathcal{P}$. Then $\Delta_{\mathcal{P}}^G$ is a simplicial complex. When G is a complete graph, we may simply write $\Delta_{\mathcal{P}}$ instead of $\Delta_{\mathcal{P}}^G$ and in this case the number of vertices is always assumed to be n.

Let $M^G(\mathcal{P}, d, k)$ be the binary matrix whose rows are indexed by all d-subgraphs of G having property $\mathcal{P}$ and whose columns are indexed by all k-subgraphs of G having property $\mathcal{P}$. In addition, $M^G(\mathcal{P}, d, k)$ has an 1-entry at cell (i, j) if and only if the ith d-subgraph is a subgraph of the jth k-subgraph, that is, $M^G(\mathcal{P}, d, k) = M(\Delta_{\mathcal{P}}^G, d, k)$. By noting that a decreasing property induces a simplicial complex, the following results follow directly from Theorems 4.2.1 and 4.2.4.

Theorem 4.3.1 *Let $\mathcal{P}$ be a monotone decreasing graph property and G a graph with n vertices. If $1 \le d < k \le n$, then $M^G(\mathcal{P}, d, k)$ is a d-disjunct matrix.*

Theorem 4.3.2 *Let $\mathcal{P}$ be a monotone decreasing graph property and G a graph with n vertices. Suppose that for any two distinct k-subgraphs A and A' of G, in $\mathcal{P}$, A contains at least two edges not in A'.*

(a) If $1 \le d < k$, then $M^G(\mathcal{P}, d, k)$ is $(d; d+1)$-disjunct.

(b) If $2 \le d < k \le 2^d$, then $M(\Delta, d, k)$ is $(d; k)$-disjunct.

For example, the family of all matchings in a fixed graph form a decreasing graph property $\mathcal{M}$. By Theorem 4.3.1, one can easily verify the following result is an extension of a result obtained by Ngo and Du [11].

Theorem 4.3.3 *Let K_{2m} be a complete graph on $2m$ vertices. Then $M^{K_{2m}}(\mathcal{M}, d, k)$ is $(d; d+1)$-disjunct.*

Given a sample of n items with d positive ones, how do we choose $M^{K_{2m}}(\mathcal{M}, d, k)$? To answer this question, Ngo and Du made the following study.

Let $g(m, k)$ be the number of k-matchings in K_{2m}. Then $M^{K_{2m}}(\mathcal{M}, d, k)$ has $g(m, d)$ rows and $g(m, k)$ columns. To compare $g(m, d)$ with $g(m, k)$, they showed the following.

Theorem 4.3.4 $g(m,k) = \binom{2m}{2k}\frac{(2k)!}{2^k k!}$. *Moreover, the sequence* $g(m,k)$ *is unimodal and peaks at* $k = \lfloor m+1-\sqrt{\frac{m+1}{2}}\rfloor$.

Proof. First, there are $\binom{2m}{2k}$ possibilities for the vertex set of a k-matching since each vertex set is a $2k$-subset. For each fixed vertex set, each matching corresponds to a partition which divides $2k$ vertices into k disjoint pairs. There are clearly $\frac{(2k)!}{2^k k!}$ such partitions. Therefore, $g(m,k) = \binom{2m}{2k}\frac{(2k)!}{2^k k!}$.

Now, look at

$$\frac{g(m,k+1)}{g(m,k)} = \frac{(2m-2k)(2m-2k-1)}{2(k+1)}.$$

To have

$$\frac{g(m,k+1)}{g(m,k)} \geq 1,$$

it suffices to have

$$2k^2 - 4mk + 2m^2 - m - 1 \geq 0,$$

that is,

$$k \leq m - \sqrt{\frac{m+1}{2}}.$$

This means that for $k \leq m - \sqrt{\frac{m+1}{2}}$, $g(m,k+1) \geq g(m,k)$ and for $k > \sqrt{\frac{m+1}{2}}$, $g(m,k+1) < g(m,k)$. Therefore, $g(m,k)$ is unimodal and peaks at $k = \lfloor m+1-\sqrt{\frac{m+1}{2}}\rfloor$. □

Now, to reduce the number of tests, for a given sample with n items and d positive ones, one should first choose m satisfying

$$n \leq \binom{m}{m+1-\sqrt{\frac{m+1}{2}}}$$

and

$$d \leq m+1-\sqrt{\frac{m+1}{2}}.$$

Then, set

$$k = m+1-\sqrt{\frac{m+1}{2}}.$$

4.4 Partial Order of Linear Spaces over a Finite Field

Let q be a prime power and $GF(q)$ a finite field of order q. Consider the m-dimensional linear space on $GF(q)$. Ngo and Du [11] gave the following construction for d-disjunct matrix $M_q(m,d,k)$.

Let $f_1, \ldots, f_t$ be a linear ordering of all d-dimensional subspaces and $g_1, \ldots, g_n$ all k-dimensional subspaces. Define matrix $M_q(m, d, k) = (a_{ij})$ by

$$a_{ij} = \begin{cases} 1 & \text{if } f_i \text{ is a subspace of } g_j \\ 0 & \text{otherwise.} \end{cases}$$

By an argument similar to the proof of Theorem 4.2.1, one can easily show the following.

Theorem 4.4.1 *$M_q(m, d, k)$ is d-disjunct.*

Let $\begin{bmatrix} m \\ k \end{bmatrix}_q$ denote the number of k-dimensional subspaces in an m-dimensional linear space over $GF(q)$. Then $M_q(m, d, k)$ has $\begin{bmatrix} m \\ d \end{bmatrix}_q$ rows and $\begin{bmatrix} m \\ k \end{bmatrix}_q$ columns. Therefore, the following result would be helpful to determine m and k for a given sample of n items with d positive ones.

Proposition 4.4.2

$$\begin{bmatrix} m \\ k \end{bmatrix}_q = \frac{(q^m - 1)(q^{m-1} - 1) \cdots (q^{m-k+1} - 1)}{(q^k - 1)(q^{k-1} - 1) \cdots (q - 1)}$$

and moreover, it is unimodal with its peak at $k = \lfloor m/2 \rfloor$.

Proof. The formula about the number of k-dimensional subspaces can be found in Chapter 24 of [13]. By the formula,

$$\frac{\begin{bmatrix} m \\ k+1 \end{bmatrix}_q}{\begin{bmatrix} m \\ k \end{bmatrix}_q} = \frac{q^{m-k} - 1}{q^{k+1} - 1}.$$

Hence,

$$\begin{bmatrix} m \\ k+1 \end{bmatrix}_q > \begin{bmatrix} m \\ k \end{bmatrix}_q$$

if and only if $m - k > k + 1$, that is, $k \leq \lfloor m/2 \rfloor - 1$. Therefore, it reaches its maximum at $k = \lfloor m/2 \rfloor$. □

Given a sample of n items with d positives, by the above result one would first choose m to satisfy

$$n \leq \begin{bmatrix} m \\ \lfloor m/2 \rfloor \end{bmatrix}_q$$

and

$$d \leq \lfloor m/2 \rfloor.$$

Then, set $k = \lfloor m/2 \rfloor$.

Ngo nad Du also calculated the row-weight and the column-weight of $M_q(m,d,k)$.

Proposition 4.4.3 *$M_q(m,d,k)$ has row-weight $\begin{bmatrix} m-d \\ k-d \end{bmatrix}_q$ and column-weight $\begin{bmatrix} k \\ d \end{bmatrix}_q$.*

Proof. Clearly, the weight of each column is the number of d-dimensional subspaces in a k-dimensional subspace, i.e., $\begin{bmatrix} k \\ d \end{bmatrix}_q$. The weight $w(R)$ of a row R is the number of k-dimensional subspaces containing R. Let $R^\perp$ be the complementary orthogonal subspace of R in the m-dimensional linear space $FG(q)^m$. Then $R^\perp$ is an $(m-d)$-dimensional subspace. For each k-dimensional subspace K containing R, let $R_K^\perp$ be the complement subspace of R in K. Then $R_K^\perp$ is a $(k-d)$-dimensional subspace in $R^\perp$. Conversely, each $(k-d)$-dimensional subspace of $R^\perp$ together with R spans a k-dimensional subspace in $FG(q)^m$. Therefore, $w(R)$ equals the number of $(k-d)$-dimensional subspaces in the $(m-d)$-dimensional subspace $R^\perp$, that is, $\begin{bmatrix} m-d \\ k-d \end{bmatrix}_q$. $\square$.

D'yachkov *et al.* [1] found a strong error-tolerant property of $M_q(m,d,k)$ as follows.

Theorem 4.4.4 *Suppose $k-r \geq 2$ and $p = q(q^{k-1}-1)/(q^{k-r}-1)$. Then $M_q(m,d,k)$ is $(d;z)$-disjunct for $1 \leq d \leq p$ and*

$$z = q^{k-r} \begin{bmatrix} k-1 \\ r-1 \end{bmatrix}_q - (d-1)q^{k-r-1} \begin{bmatrix} k-2 \\ r-1 \end{bmatrix}_q.$$

Proof. Consider $d+1$ distinct column-indices $C_0, C_1, ..., C_d$ of $M_q(m,r,k)$. (They are k-dimensional spaces.) Note that C_0 contains $\begin{bmatrix} k \\ r \end{bmatrix}$ r-dimensional subspaces. To estimate a lower bound for the number of r-dimensional subspaces in C_0 but not in $C_1, ..., C_d$, we may assume that every C_i for $i = 1, ..., d$ intersects C_0 with a $(k-1)$-dimensional subspace. Thus, every C_i for $i = 1, ..., d$ covers $\begin{bmatrix} k-1 \\ r \end{bmatrix}$ r-dimensional subspaces of C_0. Consider the fact that two distinct columns C_i and C_j for $1 \leq i < j \leq d$ have their coverage overlapping at a $(k-2)$-dimensional subspace in C_0. Thus, C_1 covers $\begin{bmatrix} k-1 \\ r \end{bmatrix}$ r-dimensional subspaces of C_0 and every C_i for $i = 2, ..., d$ covers at most $\begin{bmatrix} k-1 \\ r \end{bmatrix} - \begin{bmatrix} k-2 \\ r \end{bmatrix}$ r-dimensional subspaces of C_0 not covered by C_1. Therefore, the number of r-dimensional subspaces in C_0, but not in $C_1, ..., C_d$, is at least

$$\begin{aligned} z &= \begin{bmatrix} k \\ r \end{bmatrix}_q - \begin{bmatrix} k-1 \\ r \end{bmatrix}_q - (d-1)(\begin{bmatrix} k-1 \\ r \end{bmatrix}_q - \begin{bmatrix} k-2 \\ r \end{bmatrix}_q) \\ &= q^{k-r} \begin{bmatrix} k-1 \\ r-1 \end{bmatrix}_q - (d-1)q^{k-r-1} \begin{bmatrix} k-2 \\ r-1 \end{bmatrix}_q. \end{aligned}$$

To have $M_q(m,r,k)$ being d-disjunct, we must have $z > 0$, which yields $d \leq p$. □

D'yachkov *et al.* also showed that the z-value given in Theorem 4.4.4 is the best possible. Indeed, they proved that for $k - r \geq 2$ and $1 \leq d \leq q+1$, $M_q(m,r,k)$ is not $(d;z)$-disjunct.

4.5 Atomic Poset

Consider a Boolean Algebra lattice. Each set packing design uses all points of rank 1 to index its rows and sampled points of rank k to index its columns. Each design with a simplicial complex uses points at rank d to index its rows and points at rank k to index its columns. Could we generalize some set packing designs to simplicial complexes? There are many possible topics for further research in this direction.

The design using linear algebra on finite field is related to a lattice other than the Boolean Algebra lattice. Motivated by this observation, Ngo and Du [12] proposed the following problem: Which conditions must hold so that some two levels of the lattice can be used to construct d-disjunct matrices?

Huang and Weng [6] solved this problem with atomic poset.

A *poset* (partially ordered set) is a finite set P with a binary relation "$\leq$" satisfying the following conditions:

1. $x \leq x$
2. $x \leq y$ and $y \leq z$ imply $x \leq z$.
3. $x \leq y$ and $y \leq x$ imply $x = y$.

Here, $x = y$ means that x and y are identical. When $x \leq y$ and $x \neq y$, one write $x < y$. If no $z \in P$ exists such that $x < z < y$, then we say that y covers x. A *lattice* is a poset such that for any two elements x and y, there exists a unique element z covered by both x and y, and there exists a unique element w covering both x and y. Usually, we write $z = \min(x,y)$ and $w = \max(x,y)$. It is a *semi-lattice* if one of the two conditions is satisfied.

An element $x \in P$ is said to be *minimal* if no $y \in P$ such that $y < x$. If P has unique minimal element, then one denotes it by 0 and says that 0 is the *least* element. Every element covering 0 is called an *atom*. The set of all atoms in P is denoted by A_P.

For example, the Boolean algebra on a set S is a poset consisting of all subsets of S with partial ordering $\subseteq$. It has unique minimal element, the empty set $\emptyset$ and every singleton is an atom, vice versa.

Consider a poset $(P, \leq)$ with the least element 0. This poset is said to be *ranked* if there exists a function $P \to \{0, 1, 2, ...\}$, called a *rank function*, satisfying conditions

1. $\text{rank}(0) = 0$ and

2. y covering x implies $\text{rank}(y) - \text{rank}(x) = 1$.

Define

$$\text{rank}(P) = \max\{\text{rank}(x) \mid x \in P\}$$

and

$$P_i = \{x \mid x \in P, \text{rank}(x) = i\}.$$

Then, $P_0 = \{0\}$ and $P_1 = A_p$. Note that a ranked poset is also a semi-lattice.

For example, in the Boolean algebra on a set, the rank of a subset is the number of elements in the subset.

Consider a ranked poset $(P, \leq)$ with rank D. Let $0 \leq d \leq k \leq D$. Suppose $P_d = \{x_1, x_2, \ldots, x_t\}$ and $P_k = \{y_1, y_2, \ldots, y_n\}$. Then matrix $M(P, d, k) = (m_{ij})$ is defined by

$$m_{ij} = \begin{cases} 1 & \text{if } x_i \leq y_j \\ 0 & \text{otherwise.} \end{cases}$$

Theorem 4.5.1 *Let $(P, \leq)$ be a ranked poset with rank $D \geq 1$. Suppose for any $w \in P$, subposet*

$$w^+ = \{x \mid x \in P, w \leq x\}$$

is atomic. Then for any $1 \leq d \leq k \leq D$, $M(P, d, k)$ is a d-disjunct matrix.

Proof. It suffices to show that for any subset T of P_k, with cardinality $\leq d$, and $x \in P_k - T$, there exists $y \in P_d$ such that $y \leq x$ and $y \not\leq z$ for all $z \in T$. The proof is by induction on d. For $d = 1$, since $P = 0^+$ is atomic, x is the least upper bound of $[0, x] \cap P_1$ and hence $z \in T$ cannot be an upper bound of $[0, x] \cap P_1$. It follows that there exists $y \in [0, x] \cap P_1$ such that $y \not\leq z$.

For $d \geq 2$, choose an element $z \in T$. By the same argument, on can obtain an element $w \in [0, x] \cap P_1$ such that $w \not\leq z$. Therefore, $w^+ \cap T$ has at most $d - 1$ elements. Note that every element in $w^+ \cap P_d$ has rank $d - 1$ in w^+. By induction hypothesis, there exists an element $y \in w^+ \cap P_d$ such that $y \leq x$ and $y \not\in u$ for all $u \in w^+ \cap T$. For $v \in T \setminus w^+$, since $w \not\leq u$, we must have $y \not\leq v$. □

Note that every simplicial complex Δ is a ranked poset. The rank of each face is its cardinality. Moreover, for any $w \in \Delta$, subposet

$$w^+ = \{x \mid x \in \Delta, w \subseteq x\}$$

is atomic. Therefore, Theorem 4.2.1 is a corollary of Theorem 4.5.1.

Huang and Weng also found several new ranked posets in quantum matroids, satisfying the condition in the theorem.

References

[1] A. G. D'yachkov, F. K. Hwang, A. J. Macula, P. A. Vilenkin and C.-W. Weng, A construction of pooling designs with some happy surprises, *J. Comput. Biol.*, to appear.

[2] F. H. Chang, H. L. Fu, F.K. Hwang and B. C. Lin, The minimum number of e-vertex-covers among hypergraphs with e edges of given ranks, preprint, 2004.

[3] D.-Z. Du and F. K. Hwang, *Combinatorial Group Testing and Its Applications*, World Scientific, 1999.

[4] H. L. Fu and F. K Hwang, A novel use of t-packing to construct d-disjunct matrices, Disc. Appl. Math., to appear.

[5] F. K. Hwang, On Macula's error-correcting pool designs, *Discrete Mathematics*, 267 (2003) 311-314.

[6] T. Huang and C.-W. Weng, Pooling spaces and non-adaptive pooling designs, *Discrete Mathematics* 282 (2004) 163-169.

[7] A. J. Macula, A simple construction of d-disjunct matrices, *Discrete Mathematics*, 162 (1996) 311-312.

[8] A. J. Macula, Error-correcting nonadaptive group testing with d^e-disjunct matrices, *Discrete Appl. Math.* 80 (1997) 217-222.

[9] A. J. Macula, Probabilistic nonadaptive group testing in the presence of errors and DNA library screening, *Annals of Combinatorics* 2 (1999) 61-69.

[10] A. J. Macula, Trivial two-stage group testing with high error rates, *Journal of Combinatorial Optimization*, 7 (2003) 361-368.

[11] H.Q. Ngo and D.-Z. Du, New constructions of non-adaptive and error-tolerance pooling designs, *Discrete Mathematics* 243 (2002) 161-170.

[12] H. Q. Ngo and D.-Z. Du, A Survey on combinatorial group testing algorithms with applications to DNA library screening, in *Discrete Mathematical Problems with Medical Applications*, DIMACS Ser. Discrete Math. Theoret. Comput. Sci., 55 (Amer. Math. Soc., Providence, RI, 2000) pp. 171-182.

[13] J. H. van Lint and R. M. Wilson, *A Course in Combinatorics*, Cambridge University Press, Cambridge, 1992.

[14] H. Park, W. Wu, X. Wu and H. G. Zhao, DNA screening, nonadaptive group testing and simplicial complex, *Journal of Combinatorial Optimization*, 7 (2003) 389-394.

[15] W. Wu, Y. Huang, X. Huang and Y. Li, On error-tolerant DNA screening, *Discrete Applied Mathematics*, to appear.

5

Random Pooling Designs and Probabilistic Analysis

In this chapter we study pooling designs when the positive clones are identified not with certainty, but with a good probability. This could happen due to either random designs are used or deterministic designs are used beyond the range of their intended applicability.

5.1 Introduction to Random Designs

In the last three chapters we introduced several methods to construct d-disjunct, d-separable and $\bar{d}$-separable matrices. A general observation is that although these methods are highly efficient when they exist, their existence is rare. On the other hand, a random design, regardless of which type, usually exists for all t and n. Of course, the price to be paid is that there is no guarantee that they will identify all clones correctly. Therefore, there might be positive clones unidentified, called *unresolved positives*, or negative clones unidentified, called *unresolved negatives.* Let $\bar{P}$ and $\bar{N}$ denote the set of unresolved positives and the set of unresolved negatives, respectively. A surprising thing is that most random designs do well in restricting $\bar{P}$ and $\bar{N}$ to small sets.

In this chapter we will study four random designs which have been studied in the literature. Let M be a $t \times n$ binary matrix.

1. *Random incidence design* (RID). Each cell in M has probability p of being 1.
2. *Random k-set design* (RkSD). Each column in M is a random k-subset of the set $[t] = \{1, 2, ..., t\}$.
3. *Random distinct k-set design* (RDkSD). RkSD except the columns are distinct.
4. *Random r-size design* (RrSD). Each row in M is a random r-subset of the set $[n] = \{1, 2, ..., n\}$.

RID was first proposed by Erdös and Renyi [6] in some search problems. Balding *et al.* [1] and Bruno *et al.* [2] proposed RkSD and showed it to be much powerful than RID. They also alluded to controlling the column intersection to avoid similar columns. Hwang [7] carried out this idea by studying RDkSD. Another motivation to study RDkSD is that some other pooling designs are special cases of RDkSD. Hwang also proposed RrSD to compare its row structure with the column structure of RkSD. Further, RrSD is suitable for the situation when the size of a pool is restricted. Note that one can also propose random distinct r-size design (RDrSD). But, since Hwang and Liu [8] found that the performance of RkSD and RDkSD are about the same, while the performance of RrSD is much worse, there is not much motivation to study RDrSD.

The basic probabilities to be computed for these four models are $P(|\bar{N}| = u)$ and $P(|\bar{P}| = v)$. In particular, we are interested in $P(|\bar{N}| = 0)$ and $P(|\bar{P}| = 0)$, the cases that all negative clones and positive clones, respectively, are identified. From $P(|\bar{N}| = u)$ and $P(|\bar{P}| = v)$ we also obtain $E(\bar{N})$ and $E(\bar{P})$, which in turn yield

$$P^- \equiv \frac{E(\bar{N})}{n-d} = \text{probability a random negative clone is unresolved, and}$$
$$P^+ \equiv \frac{E(\bar{P})}{d} = \text{probability a random positive clone is unresolved.}$$

However, since $E(\bar{N})$ and $E(\bar{P})$ are usually quite messy, it is often easier to argue for P^- and P^+ directly.

Recently, Hwang and Liu [9] proposed a general approach to compute these probabilities. This chapter will follow their results closely. Since the formulas may be applied to large parameters, one emphasis of analysis is the computation time of a formula, which can often be translated into the number of summation signs (and their ranges) in the formula.

We will also look into the problems of choosing p, k, r to minimize these probabilities. Due to the messiness of the probability functions, only limited results have been obtained.

5.2 A General Approach to Compute Probabilities of Unresolved Clones

Let M be a $t \times n$ binary matrix and D a set of positive clones. We say the rows (columns) are ID if the distribution of the number of 1-entrices are identical for all rows (columns). If furthermore, the distribution is independent between the rows (columns), then we say the rows (columns) are IID. In the random designs we study here, the rows and the columns are always ID.

Our analysis aims at the $(\bar{d}, n)$-space for mathematical simplicity. However, we do not really treat d as absolute. Thus unless specified otherwise, we assume that a negative clone is identified only by its presence in a negative test (as in a d-disjunct matrix), but never by deduction from the fact that d positive clones have been identified elsewhere.

Partition the rows of M into three parts X, Y, Z according as whether a pool intersects D at least twice, exactly once, or none. Then a negative can be identified if and only if its column intersects Z, and a positive can be identified only if its column intersects Y. Define $f(z) = P(|Z| = z)$. We can similarly define $f(S)$ for any $S \subseteq \{x, y, z\}$.

Lemma 5.2.1 *Suppose the rows are ID. Then*

$$f(z) = \binom{t}{z} \sum_{i=z}^{t} (-1)^{i-z} \binom{t-z}{i-z} P(i \text{ specified rows including } Z \text{ not intersecting } D).$$

Proof. $f(z)$ is computed by an inclusion-exclusion formula. □

Corollary 5.2.2 *Suppose the rows are ID and the columns are IID. Then*

$$f(z) = \binom{t}{z} \sum_{i=z}^{t} (-1)^{i-z} \binom{t-z}{i-z} [P(a \text{ column not intersecting the } i \text{ rows including } Z)]^d$$

where $d = |D|$.

Proof. The event "the i rows including Z not intersection D" can also be expressed as "d columns not intersecting the i rows including Z". □

When the rows are IID, $f(z)$ is much simpler.

Lemma 5.2.3 *Suppose the rows are IID. Then*

$$f(z) = \binom{t}{z} [P(a \text{ row not intersecting } D)]^z [1 - P(a \text{ row not intersecting } D)]^{t-z}.$$

Note that a negative column is unresolved if and only if it does not intersect any pool belonging to Z. For simplicity, we will say in such a case that it does not *intersect* Z.

Next, we give the probabilistic distribution function of $|\bar{N}|$.

Theorem 5.2.4 *Suppose the columns and rows are both ID. Then*

$$P(|\bar{N}| = u) = \sum_{z=0}^{t} f(z) \binom{n-d}{u} P(\bar{N} \text{ does not intersect } Z).$$

Either the column IID or the row IID will bring some simplification.

Corollary 5.2.5 *Suppose the rows are ID and the columns are IID. Then*

$$P(|\bar{N}| = u) = \sum_{z=0}^{t} f(z)\binom{n-d}{u}[P(\text{a negative column not intersecting } Z)^u \cdot P(\text{a negative column intersecting } Z)^{n-d-u}].$$

Corollary 5.2.6 *Suppose the columns are ID and the rows are IID. Let $J \subseteq \bar{N}$ be a set of j columns disjoint from D. Then*

$$\begin{aligned} &P(|\bar{N}| = u) \\ = &\sum_{z=0}^{t} f(z)\binom{n-d}{u}\sum_{j=u}^{n-d}(-1)^{j-u}\binom{n-d-u}{j-u}[P(\text{a pool in } Z \text{ does not intersect } J)]^z \\ = &\binom{n-d}{u}\sum_{j=u}^{n-d}(-1)^{j-u}\binom{n-d-u}{j-u}[P(\text{a pool is negative and does not intersect } J) \\ &+P(\text{a pool is positive})]^t. \end{aligned}$$

Note that the second expression gets rid of a sum over t. So, this expression will be used in future in reference to Corollary 5.2.6 when both are applicable. However, the first expression becomes competitive when columns are also IID since it is reduced to Corollary 5.2.5 except $f(z)$ can be expressed in the form of Lemma 5.2.3. Namely,

Theorem 5.2.7 *Suppose both columns and rows are IID. Then*

$$P(|\bar{N}| = u) = \sum_{z=0}^{t}\binom{t}{z}P(E)^z[1-P(E)]^{t-z}\binom{n-d}{u}P(E')^u[1-P(E')]^{n-d-u},$$

where E is the event that a row does not intersect D and E' is the event that a negative column does not intersect Z.

Note that the event "a pool is positive" in the second expression cannot be turned into a requirement on columns to take advantage of column IID. Since $t < n$ in all practical designs, Theorem 5.2.7 beats a formula derivable from the second expression.

A positive clone can be identified if and only if it is contained in a pool not containing any unresolved negative clone. Therefore, a positive clone is unresolved if and only if every pool containing it must contain either another positive clone or an unresolved negative clone.

Let q_S denote the number of rows in S not containing any unresolved negative. Then

$$P(|\bar{P}| = v) = \sum_{z=0}^{t}\sum_{y=0}^{t-z} f(z,y)P(|\bar{P}| = v \mid |Z| = z, |Y| = y)$$

$$= \sum_{z=0}^{t}\sum_{y=0}^{t-z} f(z,y) \sum_{u=0}^{n-d} P(|\bar{N}| = u \mid |Z| = z)P(|\bar{P}| = v \mid |Y| = y, |\bar{N}| = u)$$
$$= \sum_{z=0}^{t}\sum_{y=0}^{t-z} f(z,y) \sum_{u=0}^{n-d} P(|\bar{N}| = u \mid |Z| = z)$$
$$\cdot \sum_{q_Y=0}^{y} P(q_Y \mid |Y| = y, |\bar{N}| = u)P(|\bar{P}| = v \mid q_Y).$$

Therefore, we have

Theorem 5.2.8 *Suppose the columns and rows are both ID. Then*

$$P(|\bar{P}| = v) = \sum_{z=0}^{t}\sum_{y=0}^{t-z} f(z,y) \sum_{u=0}^{n-d} P(|\bar{N}| = u \mid |Z| = z)$$
$$\cdot \sum_{q_Y=0}^{y} P(q_Y \mid |Y| = y, |\bar{N}| = u)P(|\bar{P}| = v \mid q_Y).$$

Note that the above formula is not computable since we don't know how to compute the last term. This problem can be solved when rows are IID. The formula is also simplified since the summation of y and q_Y can be combined into one step.

Theorem 5.2.9 *Suppose the columns are ID and the rows are IID. Then*

$$P(|\bar{P}| = v)$$
$$= \sum_{z=0}^{t} f(z) \sum_{u=0}^{n-d} \binom{n-d}{u} \sum_{j=u}^{n-d} (-1)^{j-u} \binom{n-d-u}{j-u} P(a \text{ pool in } Z \text{ does not intersects } J)^z$$
$$\sum_{q_{X\cup Y}=0}^{t-z} \binom{t-z}{q_{X\cup Y}} P(E)^{q_{X\cup Y}} [1-P(E)]^{t-z-q_{X\cup Y}}$$
$$\binom{d}{v} \sum_{j=0}^{d-v} (-1)^j \binom{d-v}{d-v-j} \left(\frac{d-v-j}{d}\right)^{q_{X\cup Y}},$$

where E is the event that a row in $X \cup Y$ contains a single positive clone but no unresolved negative.

Proof. $q_{X\cup Y}$ represents the number of rows in $X \cup Y$ containing a single positive and no unresolved negative. Then $P(|\bar{P}| = v \mid q_{X\cup Y})$ can be viewed as the probability of getting v empty holes in rolling $q_{X\cup Y}$ balls into d holes.

Since Z and $X \cup Y$ are disjoint, the event "pools in Z do not intersect J" and E are independent. Hence the probability of the joint event is the product. □

Corollary 5.2.10 *If columns are also IID, then the sum on j can be replaced by $P(E')^u[1-P(E')]^{n-d-u}$.*

We next discuss the computation of P^- and P^+. For convenience, in computing P^-, the negative clone whose resolvability is in concern will be denoted by C. In computing P^+, D_1 is the positive clone in concern and D_j is a positive clone other than D_1. We first give a general formula for computing P^-.

Theorem 5.2.11 $P^-(C) = \sum_{z=0}^{t} f(z)P(Z \text{ does not contain } C)$.

Corollary 5.2.12 *P^- is independent of n if the columns are independent.*

Proof. $f(z)$ is determined by the columns of D, hence independent of n. □

Corollary 5.2.13 *Suppose the rows are independent. Then*

$$P^-(C) = \sum_{z=0}^{t} f(z)P(\text{a pool of } Z \text{ does not intersect } C)^z.$$

But we can do better by combining the computations of the probability of z and of the property.

Theorem 5.2.14 *Suppose the rows are independent. Then*

$$P^-(C) = [1 - P(\text{a pool intersects } C, \text{ but none of } D)]^t.$$

Even without the row independence, we can interpret Theorem 5.2.11 in a way such that there is no need to compute $f(z)$.

Theorem 5.2.15 *Suppose C has weight k. Then*

$$P^-(C) = \sum_{i=0}^{k} (-1)^i \binom{k}{i} P(\text{none of the } i \text{ specified rows intersects } D).$$

Proof. The inclusion-exclusion formula computes the probability that at least one of the k rows does not intersect D. □

Corollary 5.2.16 *Suppose the columns are IID. Then*

$$P^-(C) = \sum_{i=0}^{k} (-1)^i \binom{k}{i} P(\text{none of the } i \text{ specified rows intersects } D_1)^d.$$

To compute P^+, let Y_1 denote the subset of Y intersecting D_1, but not any other D_j. Define

$$f(z, y_1) = P(|Z| = z, |Y_1| = y_1).$$

Theorem 5.2.17 *Suppose the rows are ID. Then*

$$P^+(D_1) = \sum_{z=0}^{t}\sum_{y_1=0}^{t-z} f(z,y_1)\sum_{i=0}^{y_1}(-1)^i\binom{y_1}{i}P(\textit{all negative clones either appearing in } Z \textit{ or not appearing in } Z \textit{ and the } i \textit{ specified rows of } Y_1).$$

Proof. The $i=0$ term is clearly 1. The remaining term (changing signs) gives the probability that at least one in Y_1 does not contain an unresolved negative (hence D_1 is identified). □

Corollary 5.2.18 *Suppose the rows are ID and the columns are IID. Then*

$$P^+(D_1) = \sum_{z=0}^{t}\sum_{y_1=0}^{t-z} f(z,y_1)\sum_{i=0}^{y_1}(-1)^i\binom{y}{i}[P(\textit{a negative clone appears in } Z) + P(\textit{a negative clone appears neither in } Z, \textit{ nor in the } i \textit{ specified rows of } Y_1)]^{n-d}.$$

With IID rows, both expressions of Corollary 5.2.6 can be used. So we choose the second one.

Theorem 5.2.19 *Suppose the columns are ID and the rows are IID. Then*

$$P^+(D_1) = \sum_{u=0}^{n-d}\binom{n-d}{u}\sum_{j=u}^{n-d}(-1)^{j-u}\binom{n-d-u}{j-u}[P(\textit{a pool is positive and does not identify } D_1 \textit{ given } J) + P(\textit{a pool is negative and does not intersect } J)]^t.$$

Proof.

$$\begin{aligned}
&P^+(D_1)\\
&= \sum_{u=0}^{n-d} P(|\bar N| = u)P(\text{unresolved } D_1 \mid |\bar N| = u)\\
&= \sum_{u=0}^{n-d}\binom{n-d}{u}\sum_{j=u}^{n-d}(-1)^{j-u}\binom{n-d-u}{j-u}\\
&\quad\cdot[P(\text{a pool is negative and does not intersect } J) + P(\text{a pool is positive})]^t\\
&\quad\cdot P(\text{unresolved } D_1 \mid |J| = j) \qquad \text{by Corollary 5.2.6}\\
&= \sum_{u=0}^{n-d}\binom{n-d}{u}\sum_{j=u}^{n-d}(-1)^{j-u}\binom{n-d-u}{j-u}\\
&\quad\cdot[P(\text{a pool is negative and does not intersect } J \mid |\bar N| = u)\\
&\quad+P(\text{a pool is positive and not identifying } D_1 \mid |J| = j)]^t.
\end{aligned}$$

□

Finally, we have

Theorem 5.2.20 *Suppose both columns and rows are IID. Then*

$$\begin{aligned} & P^+(D_1) \\ = & \sum_{z=0}^{t} \binom{t}{z} P(E)^z [1-P(E)]^{t-z} \sum_{u=0}^{n-d} \binom{n-d}{u} P(E')^u [1-P(E')]^{n-d-u} \\ & \cdot \frac{1 - P(\text{a row in } T \setminus Z \text{ contains } D_1 \text{ but no other } D_j \text{ or unresolved negative})}{[1 - P(\text{a row not intersecting } D)]^{t-z}}, \end{aligned}$$

where E and E' are as defined in Theorem 5.2.7.

5.3 Random Incidence Designs

Let M be a $t \times n$ RID. Note that both rows and columns are IID. Using the row independence, it is easily obtained:

Lemma 5.3.1 $f(z) = \binom{t}{z}(1-p)^{dz}[1-(1-p)^d]^{t-z}$.

Hwang [7] used Theorem 5.2.7 to obtain an $O(t)$ time formula.

Theorem 5.3.2

$$P(|\bar{N}| = u) = \sum_{z=0}^{t} \binom{t}{z} (1-p)^{dz}[1-(1-p)^d]^{t-z} \binom{n-d}{u} (1-p)^{zu}[1-(1-p)^z]^{n-d-u}.$$

Proof. Follows immediately from Corollary 5.2.5 and Lemma 5.3.1. □

The special case $u = 0$ was first given by Balding *et al.* [1].

Corollary 5.3.3 $P(|\bar{N}| = 0) = \sum_{z=0}^{t} \binom{t}{z}(1-p)^{dz}[1-(1-p)^d]^{t-z}[1-(1-p)^z]^{n-d}$.

Note that using the row independence of Corollary 5.2.6 will require more computation.

Corollary 5.3.4 $E(|\bar{N}|) = (n-d)[1-p(1-p)^d]^t$.

Proof.

$$\begin{aligned} E(|\bar{N}|) &= \sum_{z=0}^{t} \binom{t}{z} (1-p)^{dz}[1-(1-p)^d]^{t-z} \sum_{u=0}^{n-d} u \binom{n-d}{u} (1-p)^{zu}[1-(1-p)^z]^{n-d-u} \\ &= \sum_{z=0}^{t} \binom{t}{z} (1-p)^{dz}[1-(1-p)^d]^{t-z}(n-d)(1-p)^z \\ &= (n-d)[\sum_{z=0}^{t} \binom{t}{z} (1-p)^{(d+1)z}[1-(1-p)^d]^{t-z}] \\ &= (n-d)[(1-p)^{d+1} + 1 - (1-p)^d]^t \\ &= (n-d)[1-p(1-p)^d]^t. \end{aligned}$$

□

Corollary 5.3.5 $P^- = [1 - p(1-p)^d]^t$.

Note that Corollary 5.3.5 can also be argued directly from Theorem 5.2.14 by noting that $p(1-p)^d$ is the probability that a row contains C but none of D. Then Corollary 5.3.4 can be obtained by multiplying by $(n-d)$. We did it the hard way just for demonstration purpose.

Let p_*^- minimize P^- (or $E(\bar{N})$). Bruno *et al.* [2] gave

Theorem 5.3.6 $p_*^- = 1/(d+1)$.

Proof. Clearly, to maximize P^- is to minimize $p(1-p)^d$. Set $\frac{d}{dp}p(1-p)^d = (1-p)^d - pd(1-p)^{d-1} = 0$. We obtain $p_*^- = 1/(d+1)$. □

No analytic solution of p to minimize $P(|\bar{N}| = 0)$ has been given.

The corresponding probabilities of unresolved positive are considerably messier.

Theorem 5.3.7

$$\begin{aligned} P(|\bar{P}| = v) &= \sum_{z=0}^{t} \binom{t}{z} (1-p)^{dz} [1-(1-p)^d]^{t-z} \\ &\cdot \sum_{u=0}^{n-d} \binom{n-d}{u} (1-p)^{zu} [1-(1-p)^z]^{n-d-u} \\ &\cdot \sum_{q=0}^{t-z} \binom{t-z}{q} \left[\frac{dp(1-p)^{d-1+u}}{1-(1-p)^d}\right]^q \left[\frac{1-dp(1-p)^{d-1+u}}{1-(1-p)^d}\right]^{t-z-q} \\ &\cdot \binom{d}{v} d^{-q} \sum_{j=0}^{d-v} (-1)^j \binom{d-v}{d-v-j} (d-v-j)^q. \end{aligned}$$

Proof. By Corollary 5.2.10. □

The quantity in Theorem 5.3.7 can be computed in $O(t^2nd)$ time.

As $P(|\bar{P}| = v)$ is unwieldy to maneuver, it is desirable to derive P^+ and $E(\bar{P})$ independently. We give several such derivations and compare their time complexities. First a lemma:

Lemma 5.3.8 $f(z, y_1) = \binom{t}{z, y_1} (1-p)^{dz} [p(1-p)^{d-1}]^{y_1} [1-(1-p)^{d-1}]^{t-z-y_1}$.

Proof. A pool is not in $Z \cup Y_1$ if and only if it contains a positive clone other than D_1. □

We can use the column IID formula to compute P^+.

Theorem 5.3.9

$$P^+ = \sum_{z=0}^{t}\sum_{y_1=0}^{t-z}\binom{t}{z, y_1}(1-p)^{dz}[p(1-p)^{d-1}]^{y_1}[1-(1-p)^{d-1}]^{t-z-y_1}$$
$$\cdot\sum_{i=0}^{y_1}(-1)^i\binom{y_1}{i}[1-(1-p)^z+(1-p)^{z+i}]^{n-d}.$$

Proof. $1-(1-p)^z$ is the probability that a negative clone appears in Z, and $(1-p)^{z+i}$ is the probability that a negative clone does not appear in Z or in the i specified rows of Y_1. Theorem 5.3.9 follows immediately from Corollary 5.2.18. □

Note that P^+ in Theorem 5.3.9 can be computed in $O(t^3)$ time. Alternatively, we can use the row IID formula.

Theorem 5.3.10

$$P^+ = \sum_{u=0}^{n-d}\binom{n-d}{u}\sum_{j=u}^{n-d}(-1)^{j-u}\binom{n-d-u}{j-u}[1-(1-p)^d-p(1-p)^{d-1+u}+(1-p)^{d+j}]^t.$$

Proof. $1-(1-p)^d$ is the probability that a pool contains a positive clone, hence is positive. In a positive pool, D_1 is identified if and only if it is the only positive clone in the pool and no unresolved negative is in the pool. The probability of this is $p(1-p)^{d-1+u}$ given there are u negative pools. Therefore $1-(1-p)^d-p(1-p)^{d-1+u}$ is the probability that a pool is positive but not identifying D_1 given $\bar{N}=u$.

On the other hand, $(1-p)^d$ is the probability that a pool is negative and $(1-p)^j$ is the probability that it does not contain the j specified negative clones including $\bar{N}$. Hence $(1-p)^{d+j}$ is the probability that both events happen. Theorem 5.3.10 follows from Theorem 5.2.19 immediately. □

P^+ in Theorem 5.3.10 can be computed in $O(n^2)$ time.

Finally, we can use Theorem 5.2.20 to obtain an $O(tn)$ time formula.

Theorem 5.3.11

$$P^+ = \sum_{z=0}^{t}\binom{t}{z}(1-p)^{dz}[1-(1-p)^d]^{t-z}\sum_{u=0}^{n-d}\binom{n-d}{u}(1-p)^{zu}[1-(1-p)^z]^{n-d-u}$$
$$\cdot\left[1-\frac{p(1-p)^{d-1+u}}{1-(1-p)^d}\right]^{t-z}.$$

Proof. $p(1-p)^{d-1+u}$ is the conditional probability that a pool contains D_1 but no other D_i nor an unresolved negative. Its division by $1-(1-p)^d$ gives the same probability conditional on the pool being positive (in $X \cup Y$). Theorem 5.3.11 follows from Theorem 5.2.20 immediately. □

Corollary 5.3.12 $E(|\bar{P}|) = dP^+$.

No analytic solution of p has been given to minimize either P^+ or $P(|\bar{P}| = 0)$.

Sebö [14], following D'yachkov [5], considered the minimization of $t(d,n)$ and argued that $p = 1 - (1/2)^{1/d}$ should be chosen to maximize the entropy of a test. He proved that with such a choice the upper bound in

$$d \log n(1 + o(1)) \leq t(d,n) \leq d \log n(1 + o(1))$$

is achieved, while the lower bound is simply the information-theoretic bound.

5.4 Random k-Set Designs

The columns in RkSD are IID but the rows are only ID. For example, for $t = n = 2$ and $k = 1$, if the first row has weight 1, the second must have weight 1, too.

Lemma 5.4.1

$$f(z) = \binom{t}{z} \sum_{i=z}^{t} (-1)^{i-z} \binom{t-z}{i-z} \left[\frac{\binom{t-i}{k}}{\binom{t}{k}} \right]^d .$$

Proof. Apply Corollary 5.2.2. □

Theorem 5.4.2

$$P(|\bar{N}| = u) = \sum_{z=0}^{t} \binom{t}{z} \sum_{i=z}^{t} (-1)^{i-z} \binom{t-z}{i-z} \left[\frac{\binom{t-z}{k}}{\binom{t}{k}} \right]^d \binom{n-d}{u} \left[\frac{\binom{t-z}{k}}{\binom{t}{k}} \right]^u \left[1 - \frac{\binom{t-z}{k}}{\binom{t}{k}} \right]^{n-d-u} .$$

Proof. The probability that a negative clone does not appear in a row of Z is $\binom{t-z}{k}/\binom{t}{k}$. Theorem 5.4.2 now follows from Corollary 5.2.5 and Lemma 5.4.1 immediately. □

It is easier to argue for P^- independently than from $P(|\bar{N}| = u)$.

Theorem 5.4.3

$$P^- = \sum_{i=0}^{k} (-1)^i \binom{k}{i} \left[\binom{t-i}{k} / \binom{t}{k} \right]^d .$$

Proof. The probability that a positive clone does not appear in i of the k appearances of C is $\binom{t-i}{k}/\binom{t}{k}$. Theorem 5.4.3 follows from Corollary 5.2.16 immediately. □

Corollary 5.4.4 $E(|\bar{P}|) = (n-d)P^-$.

Let k_*^- minimize P^-. To facilitate the search of k_*^-, Hwang and Liu approximate P^- by assuming t is much larger than kd,

$$\begin{aligned} P^- &\sim \sum_{i=0}^{k}(-1)^i\binom{k}{i}(1-\frac{i}{t})^{kd} \\ &\sim \sum_{i=0}^{k}\binom{k}{i}e^{-(\frac{kd}{t})^i} \\ &= \left(1-e^{-\frac{kd}{t}}\right)^k. \end{aligned}$$

Then $k' = t\ln 2/d$ minimizes $(1-e^{-kd/t})^k$. Note that $t/k'd = 1/\ln 2$ which, although not truely fulfilling the assumption t much larger than kd, is at least larger.

Percus *et al.* [13] gave a more accurate approximation

$$\begin{aligned} P^- &\sim \left[1-\left(1-\frac{k}{t}\right)^d\right]^k - \frac{dk}{t(t-k)}\binom{k}{2}\left(1-\frac{k}{t}\right)^{2d}\left[1-\left(1-\frac{k}{t}\right)^d\right]^{k-2} \\ &+\frac{c}{2}\left[\ln\left(1-\frac{k}{t}\right)\right]^2\left[k^2\left(1-\frac{k}{t}\right)^{2d}-k\left(1-\frac{k}{t}\right)^d\right]\left[1-\left(1-\frac{k}{t}\right)^d\right]^{k-2}. \end{aligned}$$

They noted that the first term gives the probability that in each of the k appearances of C, a positive clone is present assuming the pools are independent. The second term is a correction to that untrue assumption, and the third term reflects the consequences of dispersion and discreteness of the number of positives.

Since we don't know how to compute $P(|\bar{P}| = v \mid q_Y)$, we don't have a formula for $P(|\bar{P}| = v)$. To compute P^+, we need $f(z, y_1)$.

Lemma 5.4.5

$$f(z,y_1) = \binom{t}{z,y_1}\left[\frac{\binom{t-z-y_1}{k-y_1}}{\binom{t}{k}}\right]\sum_{h=0}^{t-z-y_1}(-1)^h\binom{t-z-y_1}{h}\left[\frac{\binom{t-z-y_1-h}{k}}{\binom{t}{k}}\right]^{d-1}.$$

Proof. By definitions of z and y_1, each of the remaining $t-z-y_1$ pools must contain a D_i, $i \neq 1$. The last sum in Lemma 5.4.5 gives this probability using the inclusion-exclusion formula, where $\binom{t-z-y_1-h}{k}/\binom{t}{k}$ is the probability that D_1 does not appear in a specified set of $z+y_1+h$ pools (including the pools in $Z\cup Y_1$). Finally, D_1 must appear in the y_1 rows of Y_1. Its other $k-y_1$ appearances must not be in $Z \cup Y_1$. □

Theorem 5.4.6

$$\begin{aligned} P^+(D_1) &= \sum_{z=0}^{t}\sum_{y_1=0}^{k}\binom{t}{z,y_1}\left[\frac{\binom{t-z-y_1}{k-y_1}}{\binom{t}{k}}\right]\sum_{h=0}^{t-z-y_1}(-1)^h\binom{t-z-y_1}{h}\left[\frac{\binom{t-z-y_1-h}{k}}{\binom{t}{k}}\right]^{d-1} \\ &\cdot\sum_{i=0}^{y_1}(-1)^i\binom{y}{i}\left[\frac{\binom{t}{k}-\binom{t-z}{k}+\binom{t-z-i}{k}}{\binom{t}{k}}\right]^{n-d}. \end{aligned}$$

Proof. $\{\binom{t}{k} - \binom{t-z}{k}\}/\binom{t}{k}$ is the probability that a negative clone appears in Z. $\binom{t-z-i}{k}/\binom{t}{k}$ is the probability that a negative clone does not appear in z, or in the i specified pools of Y_1. Theorem 5.4.6 follows from Corollary 5.2.18 immediately. □

Note that P^+ in Theorem 5.4.6 can be computed in $O(t^2k^2)$ time.

Corollary 5.4.7 $E(|\bar{P}|) = dP^+$.

No analytic solution of p has been given to minimize either P^+ or $P(|\bar{P}| = 0)$.

5.5 Random r-Size Designs

The rows of RrSD are IID but the columns are only ID.

Lemma 5.5.1

$$f(z) = \binom{t}{z}\left[\binom{n-d}{r}/\binom{n}{r}\right]^z \left[1 - \binom{n-d}{r}/\binom{n}{r}\right]^{t-z}.$$

Proof. $\binom{n-d}{r}/\binom{n}{r}$ is the probability that a pool does not contain any positive clone, hence it is in Z. □

Theorem 5.5.2

$$P(|\bar{N}| = u) = \binom{n-d}{u}\sum_{j=u}^{n-d}(-1)^{j-u}\binom{n-d-u}{j-u}\left[\frac{\binom{n-d-j}{r}}{\binom{n}{r}} + 1 - \frac{\binom{n-d}{r}}{\binom{n}{r}}\right]^t.$$

Proof. $\binom{n-d-j}{r}/\binom{n}{r}$ is the probability that a pool in Z does not contain any of the j specified negative clones including the given u ones. Theorem 5.5.2 now follows from Corollary 5.2.6 immediately. □

It is simpler to derive P^- directly.

Theorem 5.5.3 $P^- = \left[1 - r\binom{n-r}{d}/\binom{n}{d+1}\right]^t$.

Proof. A pool identifies C if it contains C but none of D. There are r ways to choose C and $\binom{n-r}{d}$ ways to choose D. Theorem 5.5.3 now follows from Theorem 5.2.14 immediately. □

Let r_*^- minimize P^-. W.D. Lin (private communication) obtained

Theorem 5.5.4 $r_*^- \in \{\lceil r^* \rceil, \lfloor r^* \rfloor\}$ *where* $r^* = (n-d)/(d+1)$.

Proof. Clearly, minimizing P^- is the same as maximizing $r\binom{n-r}{d}/\binom{n}{r} \equiv g(r)$.

$$\frac{g(r+1)}{g(r)} = \frac{(r+1)\binom{n-r-1}{d}}{r\binom{n-r}{d}} = \frac{(r+1)(n-d-r)}{r(n-r)} = \left(1+\frac{1}{r}\right)\left(1-\frac{d}{n-r}\right).$$

When r increases, both factors decrease. Hence the ratio decreases in r, and maximum $g(r)$ is obtained at the two integers which flank the r^* satisfying $g(r^*+1)/g(r^*) = 1$, i.e., $r^* = (n-d)/(d+1)$. □

Note that r^* divided by the number of negative clones yields P_*^- in RID.

For the unresolved positives, we have

Theorem 5.5.5

$$\begin{aligned} P(|\bar{P}| = v) &= \sum_{z=0}^{t} \binom{t}{z} \left[\frac{\binom{n-d}{r}}{\binom{n}{r}}\right]^z \left[1 - \frac{\binom{n-d}{r}}{\binom{n}{r}}\right]^{t-z} \\ &\cdot \sum_{u=0}^{n-d} \binom{n-d}{u} \sum_{j-u}^{n-d} (-1)^{j-u} \binom{n-d-u}{j-u} \left[\frac{\binom{n-d-u}{r}}{\binom{n-d-u}{r}}\right]^z \\ &\cdot \sum_{q=0}^{t-z} \binom{t-z}{q} \left[\frac{d\binom{n-d-u}{r-1}}{\binom{n}{r} - \binom{n-d}{r}}\right]^q \left[1 - \frac{d\binom{n-d-u}{r-1}}{\binom{n}{r} - \binom{n-d}{r}}\right]^{t-z-q} \\ &\cdot \binom{d}{v} d^{-q} \sum_{j=0}^{d-v} (-1)^j \binom{d-v}{d-v-j} (d-v-j)^q. \end{aligned}$$

Proof. Theorem 5.5.5 follows from Corollary 5.2.9. We will only comment on the term $P(q)$ as the other terms have been obtained before.

$\binom{n-d-u}{r-1}$ is the number of ways of choosing a (positive) pool containing D_1 but no other D_j or any unresolved negative. d times the quantity counts the number of ways of choosing a single positive clone (not necessarily D_1), but no unresolved negatives. $\binom{n}{r} - \binom{n-d}{r}$ is the number of ways of choosing a positive row. So the ratio

$$\binom{n-d-z}{r-1} \Big/ \left[\binom{n}{r} - \binom{n-d}{r}\right]$$

gives the conditional probability that a positive pool contains a single positive clone and no unresolved negative (hence the positive clone is identified). □

Again, we derive P^+ independently, which requires $f(z, y_1)$.

Lemma 5.5.6

$$f(z, y_1) = \binom{t}{z, y_1} \left[\frac{\binom{n-d}{r}}{\binom{n}{r}}\right]^z \left[\frac{\binom{n-d-1}{r-1}}{\binom{n}{r}}\right]^{y_1} \left[1 - \frac{\binom{n-d}{r} + \binom{n-d-1}{r-1}}{\binom{n}{r}}\right]^{t-z-y_1}.$$

Theorem 5.5.7

$$P^+(D_1) = \sum_{u=0}^{n-d} \binom{n-d}{u} \sum_{j=u}^{n-d} (-1)^{j-u} \binom{n-d-u}{j-u} \left[\frac{\binom{n}{r} - \binom{n-d}{r} - \binom{n-d-u}{r-1}}{\binom{n}{r}} + \frac{\binom{n-d-j}{r}}{\binom{n}{r}} \right]^t .$$

Proof. $\binom{n}{r} - \binom{n-d}{r}$ is the number of ways of choosing a positive pool. $\binom{n-d-u}{j-u}$ is the number of ways of choosing a pool containing D_1 but no other D_j or an unresolved negative (D_1 is identified). Hence

$$\binom{n}{r} - \binom{n-d}{r} - \binom{n-d-u}{r-1}$$

is the number of ways of choosing a positive pool not identifying D_1.

$\binom{n-d-j}{i}$ is the number of ways of choosing a negative pool not containing any of the j specified negative clones including $\bar{N}$. Theorem 5.5.7 now follows immediately from Theorem 5.2.19. □

P^+ in Theorem 5.5.7 can be computed in $O(n^2)$ time.

Analytic solution for optimal r to minimize either P^+ or $P(|\bar{P}| = 0)$ is not known.

5.6 Random Distinct k-Set Designs

RDkSD is neither column-independent nor row-independent. Hence the computation of the probabilities of unresolved clones poses both a challenge but also an opportunity to expand the formulas beyond the independence threshold.

Lemma 5.6.1

$$f(z) = \binom{t}{z} \sum_{i=z}^{t} (-1)^{i-z} \binom{t-z}{i-z} \binom{\binom{t-i}{k}}{d} \Big/ \binom{\binom{t}{k}}{d} .$$

Proof. All k appearances of a positive clone must be outside of Z. There are $\binom{t-z}{k}$ such distinct k-sets to choose d from. Since the rows are not independent, the inclusion-exclusion formula is called for. □

Theorem 5.6.2

$$\begin{aligned} P(|\bar{N}| = u) \ &= \ \sum_{z=0}^{t} \binom{t}{z} \sum_{i=z}^{t} (-1)^{i-z} \binom{t-z}{i-z} \left[\binom{\binom{t-z}{k}}{d} \Big/ \binom{\binom{t}{k}}{d} \right] \\ &\cdot \binom{\binom{t-z}{k} - d}{u} \binom{\binom{t}{k} - \binom{t-z}{k}}{n-d-u} \Big/ \binom{\binom{t}{k} - d}{n-d} . \end{aligned}$$

Proof. There are $\binom{t-z}{k}$ k-sets not intersecting Z. d of them are chosen by the positive clones. The u unresolved negatives must be chosen from the remaining ones, while the $n-d-u$ resolved negatives must be chosen from the $\binom{t}{k}-\binom{t-z}{k}$ k-sets intersecting Z. Theorem 5.6.2 now follows from Theorem 5.2.4. □

We argue for P^- independently.

Theorem 5.6.3

$$P^- = \sum_{z=0}^{t}\binom{t}{z}\sum_{i=z}^{t}(-1)^{i-z}\binom{t-z}{i-z}\left[\binom{\binom{t-z-i}{k}}{d}\Big/\binom{\binom{t}{k}}{d}\right]\left\{\binom{t-z}{k}-d\right\}.$$

Proof. If C is unresolved, C must be chosen from the $\binom{t-z}{k}-d$ k-sets not in Z and not taken by the d positive clones. □

We do not have a formula for $P(|\bar{P}|=v)$, even $f(y,z)$ seems too difficult to attempt. Hwang and Liu [8] gave formulas for P^+ and $f(z,y_1)$.

Lemma 5.6.4

$$f(z,y_1) = \binom{t}{z,y_1}\sum_{i=1}^{t}(-1)^{i-z}\binom{t-z}{i-z}\left[\frac{\binom{\binom{t-z-i}{k}}{d}}{\binom{\binom{t}{k}}{d}}\right]\binom{t-z-y_1}{k-y_1}\frac{\binom{t-z-y_1-\epsilon}{d_1}}{\binom{\binom{t-z}{k}}{d}}$$

where $\epsilon = 0$ except if $y_1 = 0$, then $\epsilon = 1$.

Proof. There are $\binom{t-z-y_1}{k-y_1}$ ways of choosing D_1. Each D_i, $i \neq 1$, must have their k appearances not intersecting $Z \cup Y_i$, and there are $\binom{t-z-y_1}{k}$ k-sets satisfying that to choose the $d-1$ D_i from. But if $y_1 = 0$, then D_1 is also chosen from the $\binom{t-z}{k}$ k-sets, hence one k-set should be subtracted. □

Theorem 5.6.5

$$\begin{aligned} P^+ \;=\; & \sum_{z=0}^{t}\sum_{y_1=0}^{k}\binom{t}{z,y_1}\sum_{i=1}^{t}(-1)^{i-z}\binom{t-z}{i-z}\left[\binom{\binom{t-z}{k}}{d}\Big/\binom{\binom{t}{k}}{d}\right]\binom{t-z-y_1}{k-y_1} \\ & \cdot\binom{\binom{t-z-y_1}{k}-\epsilon}{d-1}\Big/\binom{\binom{t-z}{k}}{d} \\ & \cdot\sum_{i=0}^{y_1}(-1)^i\binom{y_1}{i}\binom{\binom{t}{k}-\binom{t-z}{k}+\binom{t-z-i}{k}-(d-1)}{n-d}\Big/\binom{\binom{t}{k}-d}{n-d}. \end{aligned}$$

Proof. $\binom{t}{k} - \binom{t-z}{k}$ is the number of k-sets intersecting Z, while $\binom{t-z-i}{k}$ is the number of k-sets neither intersecting Z nor the i specified rows. So a k-set taken from the union of the two sets satisfies the condition in Theorem 5.2.17. But the $d-1$ D_j, $j \neq 1$, are also taken from the second set. So these $d-1$ k-sets must be subtracted before the $n-d$ negative clones can be chosen. □

No analytic solution of optimal k to minimize any of P^-, $P(|\bar{N}|=0)$, P^-, $P(|\bar{P}|=0)$ is known. D'yachkov [5] studied the case $n = \binom{t}{k}$ and gave an asymptotic formula for the error probability, when maximum likelihood decoding is used.

The computation times for formulas in this section are the same as corresponding ones in Section 5.4 except P^-. We summarize the computation times in the following table. All entries are arguments in the $O(\cdot)$ function.

	$P(\lvert\bar{N}\rvert = u)$	P^-	$P(\lvert\bar{P}\rvert = v)$	P^+
RID	t	1	t^2nd	tn
k-set	t^2	k	?	t^2k^2
r-size	n	1	t^2nd	n^2
dist. k-set	t^2	t^2	?	t^2k^2

5.7 Intersection Pooling Designs

Let $\mathcal{D} = \{D_1, ..., D_d\}$ denote the set of d positive clones, and let C denote a random negative clone. We say row i *intersects* column j if cell (i,j) is 1. Then C is identified if there exists a row i intersecting column C but no D_j, $1 \leq j \leq d$. Note that $v_i = 0$ from which we conclude that C is negative. On the other hand, D_j is identified if

(i) there exists a row i intersecting D_j, but no other member of $\mathcal{D}$, and

(ii) all other columns intersecting row i can be identified as negative by appearing in some row with a negative outcome.

Note that requirement (i) is the same as the requirement to identify C except there are only $d-1$ other positive clones in (i) instead of d. Requirement (ii) is much more messy as it involves the determination where negative clones intersecting that row are unresolved. Besides the computational complexity issue, the joint requirement of (i) and (ii) also render a random pooling design less likely to meet them.

The motivation of intersection pooling design is to separate the two requirements. Let M be a matrix with a high probability of meeting requirement (i), and M' of meeting requirement (ii). Let M_i (M_i') denote the ith row of M (M'). Then $MM' = \{M_i \text{ intersects } M_j \mid \text{ all } (i,j)\}$ has a high probability of meeting requirements (i) and (ii) jointly.

More specifically, suppose M' has distinct columns and M contains a row M_i intersecting D_j, but no other positive clones. For a column C in M, let C' (C'') denote its corresponding column in M' (MM'). Define $R(i) = \{(M_i \text{ intersects } M_i') \mid 1 \leq i' \leq t'\}$ and $V(i) = \{i' \mid (M_i \text{ intersects } M_i') \in R(i) \text{ has a positive outcome}\}$. For $D_x'' \in D'' \setminus \{D_j''\}$, D_x'' does not intersect any row in $R(i)$ since D_x does not intersect row i of M. Hence every row in $V(I)$ must intersect D'_j or it wouldn't have a positive outcome. Therefore $D'_j \supseteq V_i$. On the other hand, $D'_j \subseteq V(i)$ for any row of M' intersecting D'_j must be in $V(i)$. Thus we have $D'_j = V(i)$. All D_j satisfying $D_j = V(i) \neq \emptyset$ for some $1 \leq j \leq t$ are identified as positive. The reason to exclude $V_i = \emptyset$ is to avoid the admission of too many negative clones into CP simply due to their absences in M_i. Note that it is possible that $D'_j = V(i)$ and $D'_j = V(i')$. But then necessarily $V(i) = V(i')$. If no i exists with $D'_j = V(i)$, then D_j is unresolved. Thus D_j is identified if and only if:

M contains a row intersecting D_j but no other positive clone, and M' contains a row intersecting D_j. (5.7.1)

Note that a negative clone C can be misidentified, i.e., $C = V(i)$ for some i, if both following events (E1) and (E2) occur:

(E1) M contains a row intersecting C and a nonempty set S of positive clone.

(E2) M' contains at least two rows intersecting C, each also intersects some member of S. Further, each row in M' not intersecting C intersects none of S.

The requirement of a row containing S or member of S is to guarantee that the row has a positive column, hence in $V(i)$. Note that if the columns of M' are distinct, then it is impossible to have a single member D_j of S, satisfying the conditions in (ii) since then D_j would be identical to C.

The intersection pooling design is expected to be used as the first stage of a trivial 2-stage pooling design, which merely generates a set CP of candidates of positive items. We still compute P^+. But instead of P^-, we compute the expected number of negative items admitted into CP, since that represents extra testing cost due to the inaccuracy of the first stage.

Throughout this section we assume that columns in M' are distinct.

Theorem 5.7.1 *Suppose M is RID(p) and M' is RID(p'). Then for the (d, n) model,*

(i) $P^+ = 1 - \{1 - [1 - p(1-p)^{d-1}]^t\}[1 - (1-p')^{t'}]$.

(ii) $E^- = t\sum_{w=1}^{n} \binom{n}{w} p^w (1-p) n - w \sum_{s=2}^{d} \binom{w}{s}\binom{n-w}{d_s}\binom{n}{d}^{-1} (w-s)\{p'[1-(1-p')^s] + (1-p')^{s+1}\}^{t'}$.

Proof. (i) By (5.7.1) and noting that $\{1 - [1 - (1-p)^{d-1}]^t\}$ is the probability that at least one row in M satisfies (5.7.1), while $[1 - (1-p')^{t'}]$ is the probability that at least one row in M' intersects X_j.

$$
\begin{aligned}
P^+ &= Prob(\text{no row in } M \text{ satisfies (5.7.1)}) \\
&= \prod_{i=1}^{t} [1 - Prob(\text{row } i \text{ in } M \text{ satisfies (5.7.1)})] \\
&= [1 - Prob(\text{row } i \text{ in } M \text{ intersects } X_j \text{ but no other positive clone})]^t \\
&= [1 - p(1-p)^{d-1}]^t.
\end{aligned}
$$

(ii) Consider row M_i in M which has w 1-entries with probability $\binom{n}{w}p^w(1-p)^{n-w}$. Suppose row M_i intersects exactly s positive clones, with probability $\binom{w}{s}\binom{n-w}{d-s}\binom{n}{d}^{-1}$. Then it intersects $w-s$ negative clones. For each such negative clone C', the probability that $C' = V(i)$ is

$$\begin{aligned} Prob((E2)) &= [Prob(\text{a row intersecting } C' \text{ and a member of } S) \\ &\quad + Prob(\text{a row not intersecting } C' \text{ and no member of } S)]^{t'} \\ &= \{p'[1-(1-p')^s] + (1-p')^{s+1}\}^{t'}. \end{aligned}$$

Finally, we multiply by t since M_i is an arbitrary row in M. □

Using the standard minimization techniques, $p^0 = 1/d$ minimizes P^+ with

$$P^+ = [1 - d(d-1)^{d-1}/d^d]^t.$$

For any given fraction α, we can select t such that $P^+ \leq \alpha$. A convenient choice for M' is to set $t' = \lceil \log_2 n \rceil$, and use any n of the $2^{t'}$ binary numbers for columns. For example for $n = 6$, M' can be

$$\begin{matrix} 0 & 0 & 0 & 1 & 1 & 1 \\ 0 & 1 & 1 & 0 & 0 & 1 \\ 1 & 0 & 1 & 0 & 1 & 0 \end{matrix}$$

Then the number of pools is $t\lceil \log_2 n \rceil$.

The following result is a w-fixed version of Theorem 5.7.1.

Theorem 5.7.2 *Suppose the t rows of M is a random set of distinct w-subsets of the set $\{1, ..., n\}$ and m' is RID(p). Then for the (d, n) model.*

(i) $P^+ = \{\binom{n-1}{w} + \binom{n-1}{w-1} - \binom{n-d}{w-1}]/\binom{n}{w}\}^t$.

(ii) $E^- = t\sum_{s=2}^{d} \binom{w}{s}\binom{n-w}{d-s}\binom{n}{d}^{-1}(w-s)\{p'[1-(1-p')^s] + (pp')^{s+1}\}^{t'}$.

Proof. (i) A row in M does not satisfy (5.7.1) either if the row does not intersect the column representing D_1, with $\binom{n-1}{w}$ ways, or it does, but it also intersects some other positive clones, with $\binom{n-1}{w-1} - \binom{n-d}{w-1}$ ways.

(ii) Derived from Theorem 5.7.1 (ii) by fixing w. □

Note that to minimize P^+ is to maximize $w\binom{n-w}{d-1}$. It is straightforward to verify $w^* = (n-d+1)/d$ is the optimal choice.

Next consider the $(\bar{d}, n)$ model, i.e., we don't know the actual size of $|D|$ except $|D| \leq d$. Note that the information of $|D|$ is needed only in the determination of optimal p or w in M. So using an estimator of $|D|$ in p^0 or w^* leads to a reduction of performance but otherwise the design and decoding for the (d, n) model is still applicable.

Finally, note that we can interchange the roles of M and M', namely, we inspect the set $R(i') = \{M_i \text{ intersect } M'_i \mid 1 \leq i \leq t\}$ to obtain CP'. By combining CP and CP', we can decrease P^+.

5.8 Subset Containment Designs in Extended Use

The topic studied in this section is a hybridization of the topics studied in the last two chapters. Recall that SCD with parameters (m, k, d) is a d-disjunct matrix [11] introduced in Chapter 4. However, suppose c is the estimated number of positive clones. Since m is typically large (so that $\binom{m}{k} \geq n$), $\binom{m}{c}$ is usually too large for practical use even for moderate d. Therefore, Macula [12] proposed to use the $(m, k, 2)$ SCD even when the number of positives c is expected to be larger. Let SCD$(m, k, d : c)$ denotes such a use. He gave some examples that P^+ is still under reasonable control.

For $c > d$, the fact that SCD cannot identify positives and negatives for certain puts it into the category of designs studied in Chapter 5. Note that each column of an (m, k, d) SCD has weight $w = \binom{k}{d}$, but not all $\binom{\binom{m}{d}}{w}$-sets are columns. Hence the column space of an SCD is a subspace of the column space of RDwSD. Since Bruno *et al.* [2] showed that the RkSD is much better than RID, while the column space of the former is a subspace of the latter (in particular, with $p = 1/2$), the question that whether a sensible structured subspace (obviously, not all subspace will do) is better than the original space arises. To this end, it is of interest to compare the (m, k, d) SCD with RDkSD. Before doing that, we need to introduce how P^- and P^+ are computed for SCD$(m, k, d : c)$.

Unfortunately, currently no method is known which computes P^- and P^+ exactly for SCD$(m, k, d : c)$. Macula gave a decoding method which identifies positive clones but not necessarily all of them. The analysis of decoding is rather simple. Hwang and Liu [10] proposed a more standard decoding (as used in d-disjunct matrices) and proved that it always identifies a maximum number of positive clones if the upper bound information is not used. Further, this standard decoding is as simple to use as Macular's decoding, though P^+ is much harder to compute (P^- is the same for these two decodings).

For a special but practical set of parameters, Hwang and Liu gave exact formulas to compute P^- and P^+. It accomplishes two things:

1. By comparing with P^- and P^+ from Macula's decoding, we learn how good the latter is.

2. By comparing with RDkSD, we get some information to the proposed question whether a structured subspace is better than the original space.

Granted that the comparisons are done only on one set of parameters, the findings are still interesting since there is no reason to believe that this set of parameters biases any side to render the comparisons unrepresentative.

Let P_n^- and P_n^+ denote the probabilities of an unresolved negative and of an unresolved positive, respectively, appearing in n clones. We will still write P_n^- and P_n^+ as P^- and P^+ if $n = \binom{m}{k}$.

Emperical results show that k should be chosen small to minimize P_n^- and P_n^+. Then the requirement $\binom{m}{k} \geq n$ forces m to be not so small. Hence the number of pools $\binom{m}{d}$ would be too large except for $d = 2$. Macula recognized this and focused his study on the $d = 2$ case. We will also assume $d = 2$ in the following sections.

A $t \times n$ SCD yields a binary t-vector as outcomes of the t pools. The *outcome graph* G has $\{1, \ldots, m\}$ as vertex set and an edge (u, v) if the row labeled by $\{u, v\}$ has outcome 1. Figure 5.1 gives an example of an SCD with $m = 5$, $k = 3$, $d = 2$, $n = 10$ and the three positive clones $\{1, 2, 3\}$, $\{2, 3, 4\}$, $\{1, 4, 5\}$.

	1 2 3	1 2 4	1 2 5	1 3 4	1 3 5	1 4 5	2 3 4	2 3 5	2 4 5	3 4 5	outcome
12	1	1	1								1
13	1			1	1						1
14		1		1							1
15			1		1	1					1
23	1						1	1			1
24		1					1		1		1
25			1					1	1		0
34				1			1			1	1
35					1			1		1	0
45						1			1	1	1

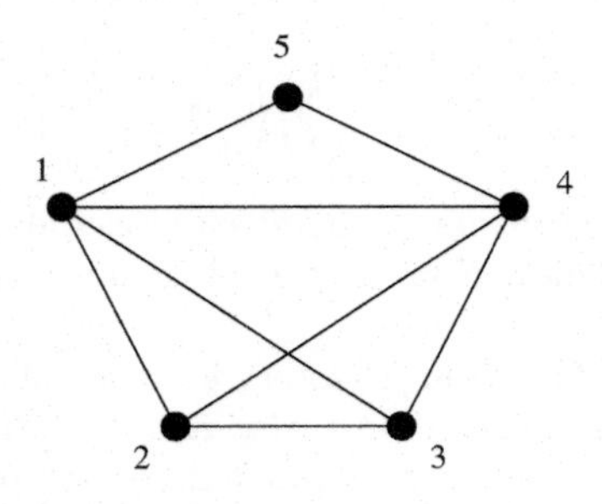

Figure 5.1: SCD(5,2,3).

Note that if a clone is positive, then every pool identifying it must be positive. This means that if a k-subset represents a positive clone, then every edge between two vertices in the k-subset must exist in the outcome graph and hence the k-subset is the vertex set of a clique in the outcome graph. Therefore, the outcome graph is actually the union of cliques corresponding to positive clones. An edge which appears in only one k-clique is called a *representative* of that clique. Hwang and Liu [10] observed

Theorem 5.8.1 *In an outcome graph, the following holds: (a) A k-clique with an edge not in the outcome graph is a resolved negative clone. Moreover, all other negative clones are unsolved.*

(b) A k-clique which has an edge-representative is a resolved positive clone. Moreover, all other positive clones are unresolved.

A vertex of a k-clique is called a *representative* if it has degree $k-1$ in the outcome graph. The following corollary was obtained by Macula [12].

Corollary 5.8.2 *In an outcome graph, the k-clique induced by vertex-representative together with its $k-1$ adjacent vertices is a positive clone.*

Proof. Let vertex v be a representative of clone C. Then v has $k-1$ edges, while each is an edge-representative of C.

Suppose to the contrary that (v, u) is not an edge-representative, i.e., $\{v, u\}$ is also an edge in another clone C', $C' \neq C$. Let w be a vertex in $V(C') \setminus V(C)$. Then $\{v, w\}$ is in G. Hence v has at least degree k, contradicting the fact that v is a vertex-representative. □

Macula gave a decoding method with this corollary, called *vertex-representative decoding*. For example, in SCD(5,3,2), suppose (1,2,3), (1,4,5) and (3,3,4) are the three positive clones. Then vertex 5 is a vertex-representative representing the positive clone $\{1,4,5\}$; however, $\{1,2,3\}$ and $\{2,3,4\}$ have no representatives and hence are not identified.

Macula also gave

$$P_n^+ = 1 - \sum_{i=1}^{k}(-1)^{i-1}\binom{k}{i}\binom{\binom{m-i}{k}}{d-1}\binom{\binom{m}{k}-1}{d-1}^{-1}$$

where i is the number of elements in the given clone-label which do not appear in the other $d-1$ positive clones. Note that P_n^+ is independent of n and hence can be written simply as P^+.

The decoding method based on Theorem 5.7.1 is called the *edge-representative decoding*, which is clearly stronger than *vertex-representative decoding*.

A generalization of the edge-representative decoding for general r, which we will call the *unique-representative decoding*, is simply the familiar decoding used for d-disjunct matrices:

1. A clone contained in a negative pool is a resolved negative clone. Moreover, all other negative clone are unresolved.

2. Remove all resolved negative clones in the design. If a clone has a unique appearance in a positive pool, then it is a resolved positive clone.

Lemma 5.8.3 *For $r = 2$, the unique-representative decoding reduces to edge-representative decoding.*

Proof. Let C denote a k-clique. Suppose C contains an edge (u, v) not in G. Then C must be negative and there exists a row $\{u, v\}$ containing C but no positive clone. So C is resolved under the unique representative decoding.

Next, suppose C is a k-clique of G and C has an edge-representative $\{u, v\}$. Then C must be positive since otherwise edge $\{u, v\}$ would not be in G. The above argument also implies that C is resolved.

Since the argument is reversible, all clones not satisfying the conditions are unresolved under the unique-representative decoding. □

Again, it is possible to use the information of d to improve the unique representative decoding. Suppose the number of resolved positive clones is d. Then all remaining clones are also resolved negative clones. If the number is less than d, then the remaining clones are unresolved clones. Also, if the outcome graph can be induced only by one set of d positives then that is it. Figure 5.1 gave such an example.

For another example, suppose $r = 2$, $d = 3$ and G is shown in Fig. 5.2. Then

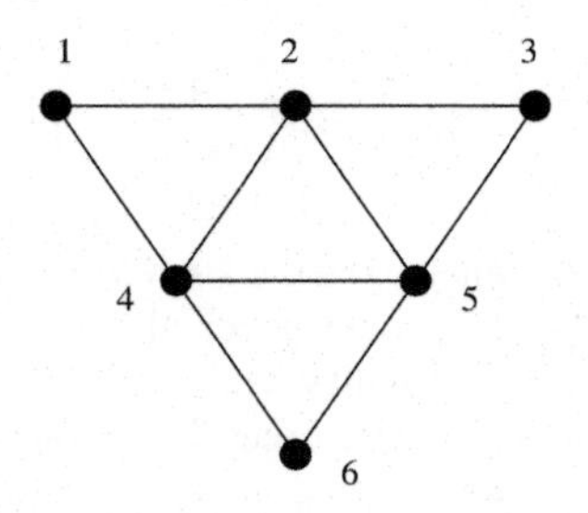

Figure 5.2: A hidden negative clone.

the three clones $\{1, 2, 4\}$, $\{2, 3, 5\}$ and $\{4, 5, 6\}$ are identified as positive by rule 2, while clone $\{2, 4, 5\}$, if it exists, is identified as negative by knowing all positives are identified.

Whenever a positive clone is resolved under the vertex-representative decoding, it is resolved under the edge-representative decoding. Further, the edge-representative decoding is at least as easy to use as the vertex-representative decoding, if not easier. Therefore, although P^+ is harder to compute under the edge-representative decoding, that shouldn't prevent it from being used as long as it guarantees better performance than its alternative.

5.9 Edge-Representative Decoding with $r = 2$ and $d = 3$

Hwang and Liu obtained an exact analysis for $d = 2$ and $c = 3$. Let C denote a clone with vertex-set C and edge set $E(C)$. Let D_1, D_2, D_3 denote the three positive clones. Define $\pi_1 = (D_1 \setminus D_2) \cap C$, $\pi_2 = (D_2 \setminus D_1) \cap C$ and $E(\pi_1, \pi_2) =$\{edges between π_1 and π_2\}.

Lemma 5.9.1 *Suppose C is negative. Then $E(C) \subseteq E(D_1) \cup E(D_2) \cup E(D_3)$ if and only if (i) there exists a pair of D_i, say D_1 and D_2 such that $C \subseteq D_1 \cup D_2$, (ii) $E(\pi_1, \pi_2) \subset E(D_3)$, (iii) $D_1 \cap D_2 \cap C \neq \emptyset$.*

Proof. Necessity. (i) Suppose that there exists a vertex $v \in C \setminus (D_1 \cup D_2)$. Note that $E(D_3)$ must contain all edges from v to $C \setminus \{v\}$, which implies $C \subseteq D_3$, an absurdity since C and D_3 are distinct k-sets.

(ii) Obvious.

(iii) Suppose $D_1 \cap D_2 \cap C = \emptyset$. By (i), $\pi_1 \cup \pi_2 \subseteq C$. By (ii), D_3 must contain C, an absurdity.

Sufficiency. An edge of C with both vertices in D_1 (D_2) is covered by D_1 (D_2). An edge of C with one vertex in D_1 and the other in D_2 is covered by D_3. By (i), there is no other edge of C. □

Theorem 5.9.2

$$P^- = \binom{N-1}{3}^{-1} \frac{1}{6} \sum_{i=1}^{k-1} \binom{k}{i} \binom{i}{k-i} \sum_{j=k-i+1}^{k-1} \binom{m-k}{i+j-k} \binom{m-k}{k-j} [\binom{m+i+j-2k}{i+j-k} - 1].$$

Proof. To cover C, D_1 and D_2 must cover all k vertices of C. Suppose D_1 covers i such vertices in $\binom{k}{i}$ ways. Then D_2 must cover all remaining $k-i$ plus at least one from $C \cap D_1$. Let j denote the cardinality of $C \cap D_2$. Then D_1 must also take $k-i$ vertices out of C in $\binom{m-k}{k-i}$ ways, and D_2 takes $k-j$ vertices in $\binom{m-k}{k-j}$ ways. Finally, D_3 must take the $k-(i+j-k)$ vertices in $\pi_1 \cup \pi_2$, hence it must take $i+j-k$ vertices from the remaining $(m-k)+(i+j-k)$ vertices not in $\pi_1 \cup \pi_2$. But one of the choices is C, which must be subtracted.

Since D_1, D_1 and D_3 can be interchanged, we divide by 6 to eliminate the repetitive counting. We also divide by the total number of ways of choosing three positive clones to obtain a probability.

Corollary 5.9.3 $P_n^- = P^-$.

Proof. P_n^- can be computed by counting the number of ways choosing 3 positive clones from N to cover C and choosing $n-4$ other negative clones, then divided by the total number of ways of choosing $n-1$ clones including 3 positive ones. Note that the number of ways to choose 3 positive clones to cover C is simply $P^-\binom{N-1}{3}$. Thus

$$P_n^- = \frac{P^-\binom{N-1}{3}\binom{N-4}{n-4}}{\binom{N-1}{n-1}\binom{n-1}{3}} = P^-.$$

□

It is more tricky to compute P_n^+. Suppose we compute P_n^+ for D_1.

Lemma 5.9.4 *In computing P_n^+, D_1 is unresolved if and only if G contains a k'-clique with $k' > k$ and $E(\pi_2, \pi_3)$ is covered by unresolved negative in the n-sample.*

Proof. For D_1 to be unresolved, necessarily $D_1 \subseteq D_2 \cup D_3$, or D_1 would have a vertex-representative. Further, neither π_2 nor π_3 can be empty. Therefore, for D_1 to be unresolved, there must exist an unresolved negative clone C to cover some edges in $E(\pi_2, \pi_3)$. Let $v \in C \setminus D_1$. Then C contains v, a vertex in π_2 and a vertex in π_3, which implies $v \in D_2 \cap D_3$. Since $D_1 \subseteq D_2 \cup D_3$, v has an edge to every vertex of D_1 in G. Therefore $v \cup D_1$ is a $(k+1)$-clique. □

Define $z = \binom{k+h}{k} - 1 - I_{h=j} - I_{h=k-i-j}$ where $I_q = 1$ if q holds and $I_q = 0$ otherwise.

Theorem 5.9.5

$$\begin{aligned}
P_n^+ \leq\ & \binom{N-1}{n-1}^{-1}\binom{n-1}{2}^{-1} \cdot \frac{1}{2} \cdot \sum_{i=0}^{k-2}\sum_{j=1}^{k-i-1}\binom{k-i}{j}\binom{m-k}{k-i-j} \cdot \\
& \sum_{h=1}^{\min\{j,k-i-j\}}\binom{k-i-j}{h}\binom{m+i+j-2k}{j-h} \cdot \\
& \{\binom{N-3}{n-3} - \binom{N-3-z}{n-3}\binom{z}{0} - \binom{N-3-z}{n-4}\left[\binom{z}{1} - \binom{\binom{i+h}{i}-1}{1}\right] \\
& -\binom{N-3-z}{n-5}\left[\binom{z}{2} - \binom{\binom{i+h}{i}-1}{2} - \binom{\binom{i+h}{i}-1}{1}\binom{z-\binom{i+h}{i}+1}{1}\right] \\
& -\frac{1}{2}\sum_{x=1}^{j-1}\sum_{y=0}^{x}\binom{j}{x}\binom{x}{y}\binom{i+h}{i+j-x}\binom{i+h}{i+x-y} \\
& -\frac{1}{2}\sum_{x=1}^{k-i-j}\sum_{y=0}^{j-1-x}\binom{k-i-j}{x}\binom{x}{y}\binom{i+h}{k-i-j}\binom{i+h}{i+x-y}\}.
\end{aligned}$$

Proof. There are $\binom{k}{i}$ ways of choosing i vertices from D_1 to appear in both D_2 and D_3. Among the remaining $k-i$ vertices of D_1, D_2 chooses j (and D_3 the remaining $k-i-j$). D_2 also chooses $k-i-j$ vertices outside of D_1, and D_3 chooses j among which h are from the $k-i-j$ vertices of $D_2 \setminus D_1$, and $j-h$ from the remaining $(m-k)-(k-i-j)$ vertices. We divide by 2 since D_2 and D_3 can be interchanged.

There are $\binom{N-3}{n-3}$ ways of choosing the other $n-3$ clones. We subtract those cases for which D_1 is resolved. We count these cases according to the number of unresolved negatives contained in the $n-3$ samples. Suppose G contains a $(k+h)$-clique, where h is the number of vertices in $D_2 \cap D_3$ outside of D_1. Then any k-clique in the $(k+h)$-clique is an unresolved negative as long as it is not a positive clone or D_1 itself. We subtract 1 since D_1 is such a clique. We also subtract $I_{h=j}$ (implying D_3 is such a clique) and $I_{h=k-i-j}$ (implying D_2 is such a clique). Therefore

$$z = \binom{k+h}{k} - 1 - I_{h=j} - I_{h=k-i-j}$$

is the number of unresolved negatives. An unresolved negative is called *universal* if it contains $\pi_2 \cup \pi_3$ (hence covering $E(\pi_2, \pi_3)$). Since its other i vertices must be taken from the $i+h$ vertices of $D_2 \cap D_3$ except the choice yielding D_1, there are $\binom{i+h}{i}$ universal unresolved negatives.

If no unresolved negative is sampled, then D_1 is resolved. If one unresolved negative U is sampled, then D_1 is resolved unless U is universal.

Suppose two unresolved negatives U are sampled, then D_1 is resolved unless one of the following subcases occurs:

(i) Both U_1 and U_2 are universal.

(ii) Exactly one of U_1 and U_2 is universal.

(iii) Both U_1 and U_2 contain π_3 and together, cover π_2 (but neither is universal). More specially, U_1 consists of the $k-i-j$ vertices of π_3, x vertices of π_2, and $i+j-x$ vertices from $D_2 \cap D_3$, while U_2 consists of the $k-i-j$ vertices of π_3, the $j-x+y$ vertices of π_1 including y of the x vertices chosen in U_1, and the $i+x-y$ vertices from $D_2 \cap D_3$. We divide by 2 since U_1 and U_2 can be interchanged.

(iv) Both U_1 and U_2 contain π_2 and together, cover π_3.

If at least three unresolved negatives are sampled, we assume D_1 is unresolved even though there are counterexamples. Therefore, the formula is an upper bound of P_n^+. □

Corollary 5.9.6

$$\begin{aligned} P^+ &= \binom{N_1}{2} \frac{1}{2} \sum_{i=1}^{k-2} \binom{k}{i} \sum_{j=1}^{k-2} \binom{k-i}{j} \binom{m-k}{k-i-j} \cdot \\ & \sum_{h=1}^{\min\{j,k-i-j\}} \binom{k-i-j}{h} \binom{m+i+j-2k}{j-h}. \end{aligned}$$

Proof. When every member of $\binom{[m]}{k}$ is taken as a clone, then all the z unresolved negatives exist. Therefore, D_1 is always unresolved as long as $h \geq 1$ (a k-clique exists in G with $k' > k$).

Let $P(|\bar{P}| = p)$ denote the probability that exactly p positive clones are unresolved under the point-representative-decoding. Macula [11] approximated $P(|\bar{P}| = 0)$ by $(P^+)^3$. Hwang and Liu [10] gave an exact solution of $P(|\bar{P}| = p)$.

Theorem 5.9.7

$$(i)\ P(|\bar{P}| = 3) = \frac{1}{6}\binom{m}{k}\sum_{i\geq\lceil\frac{k}{2}\rceil}^{k-1}\binom{k}{i}\binom{m-k}{k-i}\binom{i}{2i-k}\binom{n}{3}^{-1},$$

$$(ii)\ P(|\bar{P}| = 2) = \frac{1}{2}\binom{m}{k}\sum_{i\geq\lceil\frac{k}{2}\rceil}^{k-1}\binom{k}{i}\binom{m-k}{k-i}\left[\binom{m-2(k-i)}{2i-k} - \binom{i}{2i-k}\right]\binom{n}{3}^{-1},$$

$$(iii)\ P(|\bar{P}| = 1) = \frac{1}{2}\binom{m}{k}\sum_{i\geq\lceil\frac{k}{2}\rceil}^{k-1}\binom{k}{i}\binom{m-k}{k-i}\sum_{j=1}^{k-i-1}\binom{k-i}{j} \cdot\sum_{h=1}^{k-i-1}\binom{k-i}{h}\binom{i}{k-j-h}\binom{n}{3}^{-1}.$$

Proof. Suppose D_1, D_2, D_3 are the three positive clones. Without loss of generality, assume $|D_1 \cap D_2| = i$ for $i = 0, 1, \ldots, k-1$.

(i) Each vertex in G must lie in at least two positive clones since any vertex in a single positive clone has degree $k-1$, hence would be a representative vertex. Therefore G is a k'-clique with $k' > k$. There are $\binom{m}{k}$ ways of choosing D_1, $\binom{k}{i}\binom{m-k}{k-i}$ ways of choosing D_2 which intersects D_1 in i elements, and $\binom{i}{2i-k}$ ways of choosing $2i-k$ elements from $D_1 \cap D_2$, which, together with the $k-i$ elements from $D_1 \setminus D_2$ and the $k-i$ elements from $D_2 \setminus D_1$, yield D_3. The lower bound $\lceil\frac{k}{2}\rceil$ of i guarantees $\binom{i}{2i-k} > 0$, and the upper bound $k-1$ is due to the distinctness of x, y, z. Finally, since D_1, D_2, D_3 are interchangeable, we divide the sum by six.

(ii) Let the two positive clones with no representative be D_1 and D_2. Then D_3 must contain all vertices in $D_1 \setminus D_2$ and $D_2 \setminus D_1$, as well as an element outside of $D_1 \cup D_2$. Again, there are $\binom{m}{k}\binom{k}{i}\binom{m-k}{k-i}$ ways of choosing D_1 and D_2 with i intersections. D_3 must take the $k-i$ vertices from each of $D_1 \setminus D_2$ and $D_2 \setminus D_1$, and $2i-k$ more vertices elsewhere, excluding the case that the $2i-k$ vertices are all in $D_1 \cup D_2$ (case (i)). We divide by 2 since D_1 and D_2 are interchangeable.

(iii) Let the two clones with representatives be D_1 and D_2. Then D_3 must not contain all vertices in $D_1 \setminus D_2$, or all vertices in $D_2 \setminus D_1$. Further, $D_3 \subset D_1 \cup D_2$. □

Corollary 5.9.8 $P(|\bar{P}| = 0) = 1 - P(|\bar{P}| = 3) - P(|\bar{P}| = 2) - P(|\bar{P}| = 1)$.

Next we compute $P(|\bar{P}| = p)$ without sampling under the edge-representative-decoding.

Theorem 5.9.9

$$(i)\ P(|\bar{P}|=3) = \frac{1}{6}\binom{m}{k}\sum_{i\geq\lceil\frac{k}{2}\rceil}^{k-1}\binom{k}{i}\binom{m-k}{k-i}\binom{i}{2i-k}\binom{n}{3}^{-1},$$

$$(ii)\ P(|\bar{P}|=2) = \frac{1}{2}\binom{m}{k}\sum_{i\geq\lceil\frac{k}{2}\rceil}^{k-1}\binom{k}{i}\binom{m-k}{k-i}\left[\binom{m-2(k-i)}{2i-k}-\binom{i}{2i-k}\right]\binom{n}{3}^{-1},$$

$$(iii)\ P(|\bar{P}|=1) = \frac{1}{2}\binom{m}{k}\sum_{i\geq\lceil\frac{k}{2}\rceil}^{k-1}\binom{k}{i}\binom{m-k}{k-i}\sum_{j=0}^{i-1}\binom{i}{j} \cdot \sum_{h=1}^{\min\{k-i,i-j\}-1}\binom{k-i}{h}\binom{m+i-2k}{i-j-h}\binom{n}{3}^{-1}.$$

Proof. (i) and (ii) are the same as their counterparts in Theorem 5.9.7 since the necessary and sufficient conditions for (i) and (ii) are the same regardless of point-representative or edge-representative.

(iii) Suppose D_3 is the only unresolved positive. Then necessarily, $D_3 \subset D_1 \cup D_2$ and $D_1 \cap D_2$ in $h \geq 1$ vertex outside of D_3. Further, D_3 must not contain all vertices in $D_1 \setminus D_2$ or all vertices in $D_2 \setminus D_1$.

Suppose D_1 contains i vertices from D_3, and D_2 contains $k-i+j$ vertices from D_3, including the $k-i$ vertices not in D_1. Then D_1 also contains $k-i$ vertices outside of D_3, while D_2 contains $i-j$ vertices outside of D_3, among which h also appears in D_1. □

Corollary 5.9.10 $P(|\bar{P}|=0) = 1 - P(|\bar{P}|=3) - P(|\bar{P}|=2) - P(|\bar{P}|=1)$.

5.10 Some Trivial 2-Stage Pooling Designs

In a *trivial 2-stage pooling design*, the first stage consists of a set of pools and their testing. According to the test outcomes, a set CP of candidates of positive clones is generated. The second stage consists of individual testing of all members of CP to verify their positiveness. A trivial 2-stage pooling design differs from a general 2-stage pooling design in the special set-up of the second stage. Note that for all 1-stage pooling designs where errors may occur, we can store its output of positive clones in CP and add a second trivial stage, just as what was proposed for the intersection pooling design in Section 5.7.

Since a trivial 2-stage scheme identifies only positive clones and each is verified by individual testing, no negative clone can be misclassified. On the other hand, a positive clone can be unresolved if it is not admitted into CP at the first stage. Of course, we can reduce P^+ by enlarging CP, but that would increase the cost of testing in the second stage. So a trivial 2-stage scheme is evaluated by two criteria: P^+ and the total number of tests, $t+|CP|$, where t is the number of pools in the first stage.

Let M be an arbitrary binary matrix and let V be the outcome vector after testing all pools in M. Consider the V*-containment decoding* which identifies all columns not contained in V as negative items. Then we have a 2-stage scheme by putting all remaining items into CP, i.e., CP consists of columns contained in V. Since no positive item is removed by the first-stage decoding, $P^+ = 0$. So the only question is $t + |CP|$.

Suppose we are dealing with the $(\bar{d}, n)$ space. Of course, if M is d-disjunct, then $CP = D$ and the second stage is not needed. Chen, Hwang and Li [3] considered M with weaker properties. In particular, they used the (d, r)-disjunct notion to bound $|CP|$.

Theorem 5.10.1 *Suppose M is (d,r)-disjunct. Then $|CP| \leq d + c$.*

Corollary 5.10.2 *Suppose M is d-disjunct. Then $|CP| \leq d$.*

For the $(\bar{d}, n)$ space, De Bonis, Gasieniec and Vaccaro [4] proposed to use a $(k, d+1, n)$-selector for the first stage of a trivial 2-stage scheme. By Lemma 2.2.5, this selector is $(d, k-d-1)$-disjunct. By Theorem 5.10.1, $|CP| \leq d + k - d - 1 = k - 1$. Thus,

Lemma 5.10.3 *The DGV 2-stage scheme requires at most $t + k - 1$ tests while t is the number of rows in the $(k, d+1, n)$ selector.*

Using Lemma 5.10.3 reversely, we obtain a bound of $t_s(k, m, n)$.

Corollary 5.10.4 $t_s(k,m,n) \geq \log(\sum_{i=0}^{m-1} \binom{n}{i}) - k + 1$.

Proof. By Lemma 5.10.3, $t_s(k,m,n) + k - 1$ tests identify a sample in the $(\overline{m-1}, n)$ space which has $\sum_{i=0}^{m-1} \binom{n}{i}$ sample points. Corollary 5.10.4 follows by using the information bound. □

In the original bound De Bonis, Gasieniec and Vacarro derived, $\sum_{i=0}^{m-1} \binom{n}{i}$ was replaced by $\binom{n}{m-1}$. They observed that this bound compares favorably to a bounded of $t_s(k,m,n)$ obtainable from the bound $t(m-1, k-m, n)$ of Theorem 2.2.4.

Theorem 5.10.5 *The number of tests in the 2-stage scheme of De Bonis, Gasieniec and Vacarro is bounded by* $4ed \ln \frac{n}{2d} + d(8e+2) - 2e - 1 < 7.54d \log \frac{n}{d} + 16.21d - 2e - 1$.

Proof. Inserting the upper bound of t in Theorem 2.2.6 to Lemma 5.10.3, we obtain

$$t + k - 1 \leq \frac{ck^2}{k-d} \ln \frac{n}{k} + \frac{ek(2k-1)}{k-1} + k - 1.$$

Theorem 5.10.5 follows by replacing k by its optimal choice $k = 2d$. □

Schultz [15] first observed the ability of a d-disjunct matrix to detect whether the true number of positive items $> d$. Namely, suppose after the V-containment decoding, $n_D > d$. Then the answer to the above question must be positive.

We make use of the above property to propose a trivial 2-stage scheme. Suppose the samples are from $(\bar{d}, n)$ space. We use a d'-disjunct matrix, $d' < d$, for the first stage, and set $CP = N_D$. Similarly, we can use a d'-separable or $\bar{d}'$-separable matrix for the first stage.

We will imitate Lemma 2.2.8 to give a bound of $|CP|$ when M is d'-separable.

Theorem 5.10.6 *Suppose M is d'-separable and the problem is defined on the (d, n) space with $d > d'$. Define $x = \lfloor d/(d - d' + 1) \rfloor$. Then*

$$n_D \leq 2^{\lfloor td/x \rfloor} + d - 1.$$

Proof. Let $D = \{D_1, ..., D_d\}$. Let K denote a $(d - d' + 1)$-subset of D. Define

$$K^* = (\cup K) \setminus \cup (D \setminus K),$$

namely, K^* is a set of indices from these rows each intersecting K but not $D \setminus K$. Then each column of $N_D \setminus D$ must intersect K^* in a different subset. Suppose to the contrary that

$$C \cap K^* = C' \cap K^*$$

for two such columns C and C'. Then

$$C \cup (D \setminus K) = \cup D \setminus C \cap K^* = \cup D \setminus C' \cap K^* = C' \cup (D \setminus K),$$

contradicting the d'-separability of M.

Partition D into x disjoint $(d - d' + 1)$-subsets $K_1, ..., K_x$, except $d - x(d - d' + 1)$ columns of D may be left out. Choose K^* among $\{K_1, ..., K_x\}$ to minimize $|K^*|$. Then $|K^*| \leq \lfloor t_D/x \rfloor$. Hence K^* has only $2^{\lfloor t_D/x \rfloor} - 1$ distinct subsets (the subsets K^* is subtracted). Therefore

$$n_D \leq 2^{\lfloor t_D/x \rfloor} + d - 1.$$

Define $t_d = \max_D t_D$, we obtain Theorem 5.10.6. □

For $d' = d$, Theorem 5.10.6 is reduced to Lemma 2.2.8.

References

[1] J. Balding, W. J. Bruno, E. Knill and D.C. Torney, A comparative survey of non-adaptive pooling designs, in *Genetic Mapping and DNA Sequencing*, IMA Volumes in Mathematics and Its Applications, (Springer Verlag, 1995) pp. 133-155.

[2] D. J. Bruno, E. Knill, D. J. Balding, D. C. Bruce, N. A. Doggett, W. W. Sawhill, R. L. Stalling, C. C. Whittaker and D. C. Torney, Efficient pooling designs for library screening, *Genomics*, 26 (1995) 21-30.

[3] H. B. Chen, F. K. Hwang and Y. M. Li, Bounding the number of columns which only appear in positive pools, *Taiwanese J. Math.*, to appear.

[4] A. De Bonis, L. Gasieniec and U. Vaccaro, Optimal two-stage algorithms for group testing problems, *SIAM J. Comput.*, 34 (2005) 1253-1270.

[5] A. G. D'yachkov, On a search model of false coins, *Colloq. Math. Soc. Janos Bolyai Topics in Inform. Thy.* Keszthely, Hungary, 1975, pp. 163-170.

[6] P. Erdös and A. Renyi, On two problems of information theory, *Publ. Math. Inst. Hung. Acad. Sci.*, 8 (1963) 241-254.

[7] F. K. Hwang, Random k-set pool designs with distinct columns, *Prob. Eng. Inform. Sci.*, 14 (2000) 49-56.

[8] F. K. Hwang and Y. C. Liu, The expected number of unresolved positive clones in various random pool designs, *Prob. Eng. Inform. Sci.* 15 (2001) 57-68.

[9] F. K. Hwang and Y. C. Liu, A general approach to compute the probabilities of unresolved clones in random pooling designs, to appear in *Prob. Eng. Inform. Sci.*, 18 (2004) 161-183.

[10] F. K. Hwang and Y. C. Liu, Random pooling designs under various structures, *Journal of Combinatorial Optimization*, 7 (2003) 339-352.

[11] A. J. Macula, A simple construction of d-disjunct matrices with certain weights, *Disc. Math.*, 80 (1997) 311-312.

[12] A. J. Macula, Probabilistic nonadaptive group testing in the presence of errors and DNA library screening, *Ann. Comb.*, 1 (1999) 61-69.

[13] J. K. Percus, O. E. Percus, W. J. Bruno and D. C. Torney, Asymptotics of pooling design performance, *J. Appl. Prob.*, 36 (1999) 951-964.

[14] A. Sebö, On two random search problem, *J. Statist. Plan. Infer.*, 11 (1985) 23-31.

[15] D. J. Schultz, Topics in nonadaptive group testing, Ph.D. Thesis, Temple Univ., 1992.

6

Pooling Designs on Complexes

The complex model is a (hyper)graph-version of the group testing problem. The vertices of the hypergraph still stand for the items (or clones) to be tested, but the subjects of the screening are the edges, which can be either positive or negative, not the vertices, though a test is still on a subset of vertices, not a subset if edges. The complex model not only broadens the applicability of group testing to biological screening designs, but also brings graph theory to group testing.

6.1 Introduction

The pooling design we have discussed so far has a set of positive elements each can induce a positive effect. Sometimes it takes a set of elements combined to induce a positive effect. Therefore the set of positive elements is replaced by the set $\mathcal{D} = \{D_1, ..., D_d\}$ where each D_i, called a *positive complex*, is a subset of elements. It is usually assumed that $D_i \not\subseteq D_j$ for all $i \neq j$. Torney [23] first introduced the complex model and gave the complexes of eukaryotic DNA transcription and RNA translation as an example.

Besides its applicability to molecular biology, the complex model is interesting as a genuine generalization of the classic group testing from searching subsets to searching subgraphs. Consider a hypergraph $H(V, E)$ where the vertex-set V is the set of elements (clones, molecules) and each edge represents a candidate member of the unknown $\mathcal{D}$. Edges in $\mathcal{D}$ are referred to as *positive edges* while all other edges are *negative*. The problem is to identify all positive edges of H by edge tests. An *edge-test* is a test on a subset S of vertices with two possible outcomes: a *positive outcome* indicates that S contains an edge of $\mathcal{D}$; a *negative outcome* indicates otherwise. Note that a biological assay on a pool corresponds to an edge-test since a positive outcome is obtained as long as the pool contains one positive complex (an edge). We do not distinguish between "edge" and "complex", often honoring the usage of the quoted literature. The *rank* of an edge is the number of vertices in it. Define

(i) $H_{\bar{r}}$: rank-r graph, i.e., the maximum rank of H is r,
(ii) H_r: r-graph, i.e., every edge is of rank r,
(iii) H_r^*: complete r-graph, i.e., a k-set is an edge if and only if $k = r$,

(iv) $H_{\bar{r}}^*$: complete rank-r graph, i.e., a k-set is an edge if and only if $k \leq r$.

While the group testing (n, d) model can be represented as the edge-testing (H_1^*, d) model, it is also well known (p.211 of [6]) that a group test on a set S in the (n, d) model is equivalent to an edge test on S in the $(H_{\bar{d}}^*, 1)$ model since both tests divide the candidate set into two parts, one containing candidates which intersect with S, and the other containing candidates which do not (note that the positive outcome of one test corresponds to the negative outcome of the other). Therefore, there exists a one-to-one mapping between the classical group testing algorithms of identifying d $(\bar{d})$ positive vertices and the edge-testing algorithms of identifying a single edge in $H_{\bar{d}}^*$ $(H_{\bar{\bar{d}}}^*)$.

Let $t(H : \mathcal{D})$ denote the minimum number of tests given H and $\mathcal{D}$. If the only information available about $\mathcal{D}$ is its cardinality d, then write $t(H : d)$ instead, or $t(H : \bar{d})$ if d is an upper bound. From the above discussion, we have

Theorem 6.1.1 $t(H_{\bar{1}}^* : d) = t(H_{\bar{d}}^* : 1)$.

For $d > 1$ and $r > 1$, the edge-testing model no longer has an interpretation as a group testing model. Further, the edge-testing model encounters a difficulty not present in the group testing model. In the group testing model, once a positive vertex is found, it is removed from the set and has no impact on further testing. But in the edge-testing model, once a positive edge is found we cannot remove it since the vertices in it may still appear in other positive edges. On the other hand if a positive edge is not removed, then it may be caught repeatedly. An edge-testing algorithm has to deal with this issue.

We extend the notions of d-separable and d-disjunct matrices to the edge-testing model. For a given hypergraph H, let M be the incidence matrix of an edge-testing algorithm where rows are labeled by tests and columns by vertices. Then M is called $(H : d)$*-separable* if any two different d-sets of complexes have different outcome vectors; it is $(H : \bar{d})$*-separable* if d is just an upper bound of the cardinalities of the two sets. This implies that given an outcome vector U, there exists a unique set $\mathcal{D} = \{D_1, ..., D_j\}$, $j = d$ in the d-separable case and $j \leq d$ in the $\bar{d}$-separable case, consistent with U. In the $\bar{d}$-separable case, we assume no two positive edges can contain each other for otherwise we can't tell whether only one or both are in the positive set.

M is called $(H : d)$*-disjunct* if all edges of H not in $\mathcal{D}$, $|D| \leq d$, must appear in a pool whose outcome is negative. Therefore we can eliminate all negative edges from negative pools and what left are the D_i's. For an edge X, we say a row *contains* (or *covers*) X if it intersects every vertex (column) of X. In a matrix M, $\cap X$ denotes the set of rows each containing X. Then more formally, M is $(H : d)$-disjunct if and only if for every set of $d + 1$ edges $E = \{e_0, e_1, ..., e_d\}$, there exists a row in M which contains $e_0 \in E$, but none of $E \setminus \{e_0\}$, or

$$\cap e_0 \not\subseteq \cup_{i=1}^{d} (\cap e_i).$$

Note that for two edges $e \subset e'$, there does not exist a row covering e' but not e. Hence $(H : d)$-disjunct makes sense only if the assumption that no edge contains another edge is made, which we do. In particular, $(H_r^* : d)$-disjunctness is a valid concept. Further

Lemma 6.1.2 *M is $(H : d)$-disjunct if and only if for every set of $d + 1$ edges, $\{e_0, e_1, ..., e_d\}$, there exists a row which contains e_0 but none of columns $C_1, ..., C_d$ for some $C_i \in e_i$, $1 \leq i \leq d$.*

The relations between $(H : d)$-separable, $(H : \bar{d})$-separable and $(H : d)$-disjunct are much like in group testing. We list them in the following, some obvious ones are stated without proofs.

Theorem 6.1.3 *$(H : d)$-disjunct implies $(H : \bar{d})$-separable which imples $(H; d)$-separable.*

Theorem 6.1.4 *The $(H_r^* : d)$-separable, the $(H_r^* : \bar{d})$separable and the $(H_r^* : d)$-disjunct properties are preserved if either d is reduced to d' or r to r'.*

Theorem 6.1.5 *Let M be $(H : d)$-disjunct, and let M' be obtained from M by deleting a row. Then M is $(H : \bar{d})$-separable.*

Proof. Consider two distinct sets $\mathcal{D}, \mathcal{D}'$ of complexes. Then either $\mathcal{D} \backslash \mathcal{D}'$ or $\mathcal{D}' \backslash \mathcal{D} \neq \emptyset$. Without loss of generality, assume the former and let $D_1 \in D \setminus D'$. Suppose $U(D) = U(D')$. Then D_1 is covered by $\cup_{D_i \in \mathcal{D}} D_i = U(\mathcal{D}) = U(\mathcal{D}') = \cup_{D'_i \in \mathcal{D}'} D'_i$, contradicting the assumption that M is $(H : d)$-disjunct. Next, suppose $|U(\mathcal{D}) \setminus U(\mathcal{D}')| = 1$. Then $U(\mathcal{D}') \subset U(\mathcal{D})$ implies that $\mathcal{D}'$, hence every $\mathcal{D}'_i \in \mathcal{D}'$, is covered by $\cup_{D_i \in \mathcal{D}} D_i$, again contradicting the assumption that M is $(H : d)$-disjunct. Therefore $|U(\mathcal{D}) \setminus U(\mathcal{D}')| \geq 2$; so deleting a row would preserve the distinctness of $U(\mathcal{D})$ and $U(\mathcal{D}')$. □

Lemma 6.1.6 *The $(H : \bar{d})$-separable, the $(H : d)$-separable and the $(H : d)$-disjunct properties are preserved under adding rows or dropping columns.*

Theorem 6.1.7 *$(H : \bar{d})$-separable implies $(H : d - 1)$-disjunct.*

Proof. Let M be $(H : \bar{d})$-separable. Suppose to the contrary that M is not $(H : d-1)$-disjunct. Then there exists a d-set D of edges in H and an edge $D_1 \in D$ such that D_1 is covered by the union of the other $d - 1$ edges. Then $U(\mathcal{D}) = U(\mathcal{D} \setminus \{D_1\})$, contradicting the assumption of $(H : \bar{d})$-separable. □

While the theory of $(H : d)$-disjunct is a natural extension of d-disjunct, the construction is not since if edges can have very large ranks, then even the mere requirement that for each edge there exists a row containing it may face a large matrix. To control the size of the matrix, we need to make use of an upper bound r

of edge ranks. Thus we will often represent H by $H_{\bar{r}}$. Note that this representation does not lose generality since every H has a maximum rank.

Finally, we introduce the error-tolerant version of $(H : d)$-disjunct matrix. A binary matrix is $(H : d; z)$-disjunct if for any $d+1$ edges $e_0, e_1, ..., e_d$, there exist at least z rows each covers e_0 but none of the other e_i. $(H : d; 1)$-disjunct will be simply written as $(H : d)$-disjunct. An $(H : d; z)$-disjunct matrix can identify all positive complexes with up to $\lfloor (z-1)/2 \rfloor$ errors since there exist at least $\lceil (z+1)/2 \rceil$ $(> \lfloor (z-1)/2 \rfloor)$ negative rows covering a negative edge.

6.2 A Construction of $(H : d; z)$-Disjunct Matrix

Gao, Hwang, Thai, Wu and Znati [10] extended Theorem 3.5.2 to the edge-testing model. The idea is to construct a q-nary $(H : d; z)$-disjunct matrix and then convert it to binary, while a q-nary matrix is $(H : d, z)$-disjunct if for any $d+1$ edges $e_0, e_1, ..., e_d$, there exist at least z rows in each of which {entries of e_0} $\not\supseteq$ {entries of e_i} for all $1 \leq i \leq d$.

Consider a hypergraph $H = (V, E)$ with maximum rank r. Let $GF(q)$ be a finite field of order q. Associate each vertex $v \in V$ with a distinct polynomial p_v of degree k over $GF(q)$. Thus each edge $e \in E$ associates with a subset of polynomials, $P_e = \{p_v \mid v \in e\}$. Let T be a subset of t elements in $GF(q)$. Construct a $t \times |V|$ q-nary matrix $A_H(q, k, t)$ with rows labeled by T and columns by V. Each cell (x, v) contains an element $p_v(x)$ in $GF(q)$.

Theorem 6.2.1 *Suppose $H = H_{\bar{r}}$ and $t \geq rdk + z$. Then $A_H(q, k, t)$ is q-nary $(H : d; z)$-disjunct.*

Proof. Suppose to the contrary that no such z rows exist. Then for at least $rdk+1$ values of $x \in T$, $P_{e_0}(x) \supseteq P_{e_i}(x)$ for some $1 \leq i \leq d$. Thus there exists a fixed j, $1 \leq j \leq d$, such that $P_{e_0}(x) \supseteq P_{e_j}(x)$ for at least $rk+1$ values of $x \in T$. Since each $u \in e_j$ must have $p_u(x) = p_v(x)$ for some $v \in e_0$ for these $rk+1$ x's, there exists a $v \in e_0$ such that $p_u(x) = p_v(x)$ for at least $k+1$ x's. Since both p_u and p_v are polynomials of degree k, we have $p_u = p_v$, contradicting our assumption that p_u and p_v are distinct. □

Next, we construct a binary matrix $B_H(q, k, t)$ from $A_H(q, k, t)$. $B_H(q, k, t)$ has $|V|$ columns labeled by V. For each row x of $A_H(q, k, t)$ and each set $P_e(x)$, $e \in E$, $B_H(q, k, t)$ has a row labeled by $\langle x, P_e(x) \rangle$ which has an 1-entry in cell $(\langle x, P_e(x) \rangle, v)$ if $p_v \in P_e(x)$, and a 0-entry otherwise.

Theorem 6.2.2 *Suppose $t \geq rdk + z$. Then $B_H(q, k, t)$ is $(H : d; z)$-disjunct.*

Proof. Consider $d+1$ edges $e_0, e_1, ..., e_d$. Let x be a row in $A_H(q, k, t)$ such that $P_{e_0}(x) \not\supseteq P_{e_i}(x)$ for any $1 \leq i \leq d$. Then row $\langle x, e_0 \rangle$ in $B_H(q; k, t)$ covers e_0 but not

e_i for all $1 \leq i \leq d$. By Theorem 6.2.1, $A_H(q, z, t)$ has z such rows. Hence $B_H(q, k, t)$ is $(H : d; z)$-disjunct. □

$B_H(q, k, t)$ has $t' = \sum_{x \in T} \sum_{e \in E} |P_e(x)|$ rows. $|P_e(x)|$ can be bounded in two ways. The first is $|P_e(x)| \leq |E|$. But this is not a useful bound since that leads to a bound of $t|E|$ tests while individually testing each edge of H takes only $|E|$ tests.

Next we drive the second bound. Suppose $|\{p_v(x) \mid v \in V\}| = c_x$. Then $c_x \leq q$. Each row x in $A_H(q, k, t)$ can generate at most $\sum_{i=1}^{r} \binom{c_x}{i} \leq \sum_{i=1}^{r} \binom{q}{i} \leq \binom{q+r-1}{r}$ rows in $A_H(q, k, t)$. Thus $t' \leq t\binom{q+r-1}{r}$. Let $|V| = n$. Then $n \leq q^{k+1}$. Set q to be the minimum prime power $\geq drk + z$ and approximate it by drk, and set $t = drk + z$ also approximated by drk. Assume $n \geq q^k$. Then

$$t' \leq t\binom{q+r-1}{r} \sim drk(\frac{q}{r})^r = r(dk)^{r+1} \leq r(d\log_q n)^{r+1}.$$

Note that the number of edges can certainly reach n^r. With a more detailed analysis similar to the one in Section 3.6, we have

Theorem 6.2.3 *$B_H(q, k, t)$ is $(H : d; z)$-disjunct with at most $q(q+1)^r$ rows, where*

$$q \leq (1 + o(1))\frac{d\log_2 n}{\log_2(d\log_2 n)}.$$

Row x in $A_H(q, k, t)$ is transformed to c_x rows in $B_H(q, k, t)$ by replacing each element of $GF(q)$ with a binary vector of length c_x. View this as a $l \times q$ matrix Q. Then the requirement on Q is that if x is a row in which $P_{e_i}(x) \not\subseteq P_{e_0}(x)$ for all $1 \leq i \leq d$, or equivalently, there exists a $C_{r+i} \in e_i$ for every $1 \leq i \leq d$ such that $p_{C_{r+i}}(x) \notin P_{e_0}(x)$, then there exists a row in Q intersecting each column $C_1, ..., C_r$, but none of $C_{r+1}, ..., C_{r+d}$. Chen, Du and Hwang [2] noted that if Q is a $(d, r]$-disjunct matrix, then it certainly meets the requirement. In fact, a $(d, r; z]$-disjunct matrix can be used to provide more error-tolerance.

Li, Thai, Liu and Wu [14] considered the special case that H is an r-partite graph and $z = 1$. We give the error-tolerant version here. The construction of $A_H(q, k, t)$ and $B_H(q, k, t)$ are the same as before.

Corollary 6.2.4 *For H an r-partite graph and $t \geq dk + z$, $B_H(q, k, t)$ is $(H : d; z)$-disjunct.*

The reason that t can be reduced from $drk + z$ to $dk + z$ is the following: Suppose $e_0 = \{v_1, ..., v_r\}$ and $e_i = \{v'_1, ..., v'_r\}$ where v_j and v'_j are vertices in part j, $1 \leq j \leq r$. Then $P_{e_0} = P_{e_i}$ implies $p_{v_j}(x) = p_{v'_j}(x)$ for $1 \leq j \leq r$. Thus, if $P_{e_0}(x) = P_{e_i}(x)$ for $k+1$ values of x, then $p_{v_j} = p_{v'_j}$ for $1 \leq j \leq r$. On the other hand, this mapping between v_j and v'_j does not exist for general $H_{\bar{r}}$. Therefore $P_{e_0}(x) = P_{e_i}(x)$ must occur at least $rk+1$ times in order to force one of j to have $p_{v_j}(x) = p_{v'_j}(x)$ at least $k+1$ times.

An analysis similar to before shows that the number of tests in $B_H(q,k,t)$ is about $(d \log_q n)^{r+1}$, a saving of a factor r from the general case.

6.3 $(d, r; z]$-Disjunct Matrix

We now discuss another type of disjunct matrices which is not defined on a graph, and thus seemingly unrelated to the $(H : d; z)$-disjunct matrix. Yet we will prove a surprising result that the new disjunct matrix is equivalent to an $(H_r^* : d; z)$-disjunct matrix.

A binary $t \times n$ matrix M is called $(d, r; z]$-disjunct ($(d, r; 1]$-disjunct will be written as $(d, r]$-disjunct) if for any $d+r$ columns $C_1, ..., C_{d+r}$, there exist at least z rows each intersecting $C_1, ..., C_r$, but none of $C_{r+1}, ..., C_{r+d}$, i.e.,

$$|\cap_{i=1}^{r} C_i \setminus \cup_{j=r+1}^{r+d} C_j| \geq z.$$

Note that $n \geq d+r$ is assumed.

Chen, Du and Hwang [2] proved the following theorem for $z = 1$.

Theorem 6.3.1 *$(d, r; z]$-disjunct $= (H_{\bar{r}}^* : d; z)$-disjunct.*

Proof. Suppose that M is $(d, r; z]$-disjunct. Consider $d+1$ arbitrary edges $e_0, e_1, ..., e_d$ such that $e_i \not\subseteq e_0$ for $1 \leq i \leq d$. Let $e_0 = \{C_1, ..., C_r\}$ and select C_{r+i} from $e_i \setminus e_0$ for $1 \leq i \leq d$. Set $S = \{C_{r+i} \mid 1 \leq i \leq d\}$. Note that the cardinality of S can be less than d since $C_{r+i} = C_{r+j}$ is possible for $i \neq j$.

Since M is $(d, r; z]$-disjunct, there exist z rows intersecting $C_1, ..., C_r$ but none of $C_{r+1}, ..., C_{r+d}$, i.e., each such row covers e_0 but none of e_i, $1 \leq i \leq d$. Thus, M is $(H_r^* : d; z)$-disjunct.

We next prove $(H_{\bar{r}}^* : d)$-disjunct $\Rightarrow$ $(d, r]$-disjunct. Suppose that M is not $(d, r]$-disjunct. Then there exist $d+r$ columns $C_1, ..., C_{r+d}$ such that

$$\cap_{i=1}^{r} C_i \subseteq \cup_{i=r+1}^{r+d} C_i.$$

Set $e_0 = \{C_1, ..., C_r\}$ and $e_i = \{C_2, ..., C_r, C_{r+i}\}$ for $1 \leq i \leq d$. Then

$$\cap e_0 = (\cap\{C_2, ..., C_r\}) \cap \{C_1\} \subseteq (\cap\{C_2, ..., C_r\}) \cap (\cup_{i=r+1}^{r+d}) = \cup_{i=1}^{d} (\cap e_i).$$

Hence M is not $(H_r^*; d)$-disjunct.

To extend to the general z case, suppose M is not $(d, r; z]$-disjunct. Then there exist at most $z-1$ rows each intersecting $C_1, ..., C_r$ but none of $C_{r+1}, ..., C_{r+d}$. Delete these rows, then the remaining matrix is not $(d, r]$-disjunct, hence not $(H_r^* : d)$-disjunct as we just proved. This implies that even after bringing back the deleted rows, the matrix is still not $(H_r^* : d; z)$-disjunct. $\square$

Using Theorem 6.3.1, we can translate Theorems 6.1.3–6.1.7, some only considering the special case $H = H_r^*$, to their $(d, r; z]$ versions. Here we only quote two such translations which have been independently studied [23] in terms of $(d, r; z]$ for the case $z = 1$.

Lemma 6.3.2 *The property of $(d, r; z]$-disjunct or $(\bar{d}, r; z]$-separable or $(d, r; z]$-separable is preserved by reducing d or r.*

Note that a corollary of Theorem 6.3.1 is

Lemma 6.3.3 *$(d, r; z]$-disjunct $\Rightarrow$ $(H_{\bar{r}}^* : d; z)$-separable.*

The following was proved in [23] for $z = 1$:

Lemma 6.3.4 *$(H_{\bar{r}}^* : \bar{d}; z)$-separable $\Rightarrow$ $(d-1, r; z]$-disjunct and $(d, r-1; z]$-disjunct.*

Proof. We prove only for $z = 1$ and rely on the argument in Theorem 6.3.1 for extension to general z.

Suppose M is not $(d-1, r]$-disjunct. Then there exist $r+d-1$ columns $C_1, ..., C_{r+d-1}$ such that

$$\cap_{j=1}^{r} C_j \subseteq \cup_{j=r+1}^{r+d-1} C_j.$$

Set $e_0 = \{C_1, ..., C_r\}$ and $e_j = \{C_2, ..., C_r, C_{r+j}\}$ for $1 \leq j \leq d-1$. Let $D = \{e_j \mid 0 \leq j \leq d-1\}$ and $D' = \{e_j \mid 1 \leq j \leq d-1\}$. Then

$$\cup_{e_j \in D}(\cap e_j) = \cup_{e_j \in D'}(\cap e_j).$$

Hence M is not $(H_{\bar{r}}^* : \bar{d})$-separable.

Next suppose M is not $(d, r-1]$-disjunct. Then there exist $r+d-1$ columns $C_1, ..., C_{r+d-1}$ such that

$$\cap_{j=1}^{r-1} C_j \subseteq \cup_{j=r}^{r+d-1} C_j.$$

Set $e_0 = \{C_1, ..., C_{r-1}\}$ and $e_j = e_0 \cup \{C_{r-1+j}\}$ for $1 \leq j \leq d$. Set $D = \{e_0\}$ and $D' = \{e_1, ..., e_d\}$. Then

$$\cup_{e_j \in D}(\cap e_j) = \cap e_0 = \cup_{j=1}^{d}(\cap e_j) = \cup_{e_j \in D'}(\cap e_j).$$

Hence M is not $(H_{\bar{r}}^* : \bar{d})$-separable. □

D'yachkov, Villenkin, Macula and Torney also observed the following result for $z = 1$. The extension to general z is immediate.

Lemma 6.3.5 *Let M be $(d, r; z]$-disjunct and M' be obtained from M by interchanging 0 and 1. Then M' is $(r, d; z]$-disjunct.*

Let $t(d, r, n; z]$ denote the minimum t in a $t \times n$ $(d, r; z]$-disjunct matrix.

Corollary 6.3.6 $t(d, r, n; z] = t(r, d, n; z]$.

Stinson and Wei [21], extending results of Stinson, Wei and Zhu [22] from $z = 1$ to general z, proved the following results. The first two are obvious.

Lemma 6.3.7 *Let M be a $(d, r; z]$-disjunct matrix and let M' be obtained from M by deleting a column and all rows intersecting it. Then M' is $(d-1, r; z]$-disjunct.*

Lemma 6.3.8 *Let M be a $(d, r; z]$-disjunct matrix and let M' be obtained from M by deleting a column and all rows not intersecting it. Then M' is $(d, r-1; z]$-disjunct.*

From Lemmas 6.3.7 and 6.3.8, we have

Theorem 6.3.9 $t(d, r, n; z] \geq t(d-1, r, n-1; z] + t(d, r-1, n-1; z]$.

This recursion leads to a lower bound of $t(d, r, n; z]$. First, we show a lemma. Define

$$g(d, r, n) = \frac{\binom{d+r}{r} \log n}{\log(d+r)}.$$

Lemma 6.3.10 $g(d, r, n) \leq g(d-1, r, n-1) + g(d, r-1, n-1)$.

Proof. Using the fact that $\log x / \log(x-1)$ is decreasing for $x > 1$. □

Theorem 6.3.11 *For $d + r > 2$,*

$$t(d, r, n; z] \geq 2cg(d, r, n-1) + \frac{c(z-1)}{2}\binom{d+r}{r},$$

where c is the same constant as in Theorem 2.7.8.

Proof. The proof is by induction on $r + d$. The case $r = 1$ is by Theorem 2.6.6 and (2.6.1). The case $d = 1$ is the same as the case $r = 1$ by Corollary 6.3.6. For $d \geq 2$, $r \geq 2$,

$$\begin{aligned} t(d, r, n; z] &\geq t(d-1, r, n-1; z] + t(d, r-1, n-1; z] \\ &\geq 2cg(d-1, r, n-1) + \\ &\quad \frac{c(z-1)}{2}\binom{d+r-1}{r} + 2cg(d, r-1, n-1) + \frac{c(z-1)}{2}\binom{d+r-1}{r-1} \\ &\geq 2cg(d, r, n-1) + \frac{c(z-1)}{2}\binom{d+r}{r}. \end{aligned}$$

□

Define

$$h(d, r, n) = \binom{d+r}{r}(d+r)\log n / \log\binom{d+r}{r}.$$

Then a similar argument leads to a stronger result.

Theorem 6.3.12 *There exists an integer $n^*_{d,r}$ such that for $n \geq n^*_{d,r}$,*

$$t(d, r, n; z] \geq 0.7c \cdot h(d, r, n) + c(d-1)(z-1)/2,$$

where c is the same constant as in Theorem 2.7.8.

6.4 Constructions for $(d, r; z]$-Disjunct Matrices

D'ychkov, Villenkin, Macula and Torney [8] gave a simple construction of $(d, r]$-disjunct matrices, which we will extend to the error-tolerant case.

Theorem 6.4.1 *The $\binom{n}{k} \times n$ binary matrix where the rows consist of all k-subsets of the set $[n]$, $r \leq k \leq n-d$, is $(d, r; \binom{n-d-r}{k-r}]$-disjunct.*

Proof. For a given r-set R and a d-set D, let $K = R \cup S$ where S is a set of $k-r$ elements from $[n] \setminus (R \cup D)$. Then the row corresponding to K covers R but not intersects D. Since there are $\binom{n-d-r}{k-r}$ choices of S, z is equal to that number. □

Corollary 6.4.2 $t(d, r, n) \leq \min\{\binom{n}{d}, \binom{n}{r}\}$.

Proof. The minimum of $\binom{n}{k}$ occurs at one of two extreme points. But $\binom{n}{n-d} = \binom{n}{d}$. □

One way to view this construction is that we take all unions of r rows of an $n \times n$ identity matrix. Note that the identity matrix is a d-disjunct matrix. An attempt was made in [8] to replace this identity matrix by any d-disjunct matrix M to reduce the number of rows. However, although for each member i of R, M contains a row M_i which contains i but no member of D, and the union of M_i over all $i \in R$ covers R but not intersects D, the union may be over less than r rows as M_i and M_j could be the same. Hence the union does not correspond to a row in the matrix constructed by taking all unions of r rows of M.

Stinnson, Wei and Zhu [22] suggested a different way to obtain error-tolerance for the construction in Theorem 6.4.1, which they viewed as a $(d, r]$-disjunct matrix as originally intended in [8]. By taking z copies of each row of the constructed matrix, one obtains a $(d, r; z]$-disjunct matrix. Note that this multiplication works even if the starting matrix itself is error-tolerant. Thus

Lemma 6.4.3 $t(d, r, n; zz'] \leq \min\{zt(d, r, n; z'], z't(d, r, n; z]\}$.

In particular when applied to the construction in Theorem 6.4.1, we obtain a $z\binom{n}{k} \times n$ $(d, r; z\binom{n-d-r}{k-r}]$-disjunct matrix.

q-nary codes were used in Chapter 3 to construct $(d;z)$-disjunct matrices. A similar construction also works for our current model. M is called a q-nary $(d,r;z]$-disjunct matrix if for any two disjoint sets D and R of columns, $|D|=d$ and $|R|=r$, there exist at least z rows i such that

$$\{m_{ij} \mid j \in D\} \cap \{m_{ij} \mid j \in R\} = \emptyset.$$

Stinson and Wei [21], extending a result of [8] for $z=1$, proved

Theorem 6.4.4 *The existences of a $t_1 \times n$ q-nary $(d,r;z_1]$-disjunct matrix and $t_2 \times q$ $(d,r;z_2]$-disjunct matrix imply the existence of a $t_1t_2 \times n$ $(d,r;z_1z_2]$-disjunct matrix.*

Proof. Let M_1 and M_2 denote the above $t_1 \times n$ and $t_2 \times q$ matrices, respectively. Replace each entry y in M_1 by column y of M_2. Then for each row i in M_1 such that

$$\{m_{ij} \mid j \in R\} \cap \{m_{ij} \mid j \in S\} = \emptyset,$$

there exist at least z_2 rows after the replacement such that $\cap R \not\subseteq \cup D$. Further, for $i' \neq i$, then the two sets of z_2 rows are disjoint. Hence the constructed matrix has z_1z_2 rows satisfying $\cap R \not\subseteq \cup D$. □

Again, Theorem 6.4.3 can be viewed as a mechanism to grow a $(d,r;z]$-disjunct matrix in the number of columns as well as in the error-tolerant capability. The q-nary $(d,r;z]$-disjunct matrix then becomes the engine of this mechanism. On the other hand, we also need some $(d,r;z]$-disjunct matrix for small z, perhaps just a $(d,r]$-disjunct matrix, to serve as the input of this mechanism. We will first study the construction of these matrices, and then the construction of the q-nary matrices.

As commented in Chapter 3, the incidence matrix of a T-design with blocks as rows is not good for a d-disjunct matrix since the number of rows is not smaller than the number of columns by the Fisher inequality, hence not better than the trivial d-disjunct matrix. However, for the $(d,r]$-disjunct matrix, the T-design has a second life since we only need to beat $\min\{\binom{n}{r},\binom{n}{d}\}$. Mitchell and Piper [18] obtained

Theorem 6.4.5 *A T-$(v,k,1)$ design yields a $t \times n$ $(d,r]$-disjunct matrix with $t = \binom{v}{T}/\binom{k}{T}$, $n=v$, $d=\lambda_{T-1}-1$ and $r=T-1$, where $\lambda_{T-1}=(v-T+1)/(k-T+1)$.*

Proof. For any set S of $T-1$ columns, there exist λ_{T-1} rows covering S. Let $\cap S = \{R_{i_j} \mid 1 \le j \le \lambda_{T-1}\}$ denote these rows. Then $(R_{i_x} \setminus S) \cap (R_{i_y} \setminus S) = \emptyset$ for $x \neq y$. Hence no column can intersect more than one such row, and it takes the union of at least λ_{T-1} columns to contain $\cap S$. □

Stinson [20] observed that a $3-(q^2+1,q+1,1)$ design (an *inverse plane*) always exists for a prime power q.

Corollary 6.4.6 *A $q(q^2+1) \times (q^2+1)$ $(q, 2]$-disjunct matrix exists for a prime power q.*

A T-design is called *super-simple* if every pair of blocks intersect in at most t elements. Kim and Lebedev [13] proved

Theorem 6.4.7 *A super-simple T-(v, k, λ) design yields a $t \times n$ $(d, r]$-disjunct matrix where*

$$t = \lambda \binom{v}{T} / \binom{k}{T}, n = v, d = \lambda - 1 \text{ and } r = T.$$

Proof. For any set S of T columns, there exist exactly λ rows covering it. Let $R_{i_1}, \ldots, R_{i_\lambda}$ denote those rows. Then $(R_{i_x} \setminus S) \cap (R_{i_y} \setminus S) = \emptyset$, $1 \le x < y \le \lambda$, for otherwise the two blocks (rows) i_x and i_y intersect in more than T elements. Thus a column not in S can intersect at most one such row, and it takes at least λ columns to cover $\cap S = \sum_{x=1}^{\lambda} R_{i_x}$. □

Next, we study the construction of q-nary $(d, r; z]$-disjunct matrices. First, consider the $z = 1$ case.

The MDS code was introduced in Chapter 3 to construct q-nary disjunct matrix. Sagalovich [19] gave the following result.

Lemma 6.4.8 *Any MDS code which has parameters (q, k, t), where $t \ge dr(k-1)+1$ and $q^k \ge d + r$, yields a $t \times q^k$ q-nary $(d, r]$-disjunct matrix.*

[8] used the Reed-Soloman code as an MDS code. Note that for any integer $k \ge 2$ and a prime power $q > k - 1$, there exists a $(q, k, q + 1)$ Reed-Soloman code.

Theorem 6.4.9 *Suppose $q \ge dr(k-1) + 1$. Then*

$$t(d, r, q^k] \le t(d, r, q][dr(k-1) + 1].$$

Proof. Denote by M the matrix obtained from the Reed-Soloman code by removing any $q - dr(k-1)$ rows. Then M is still a MDS code with parameters $(q, k, dr(k-1)+1)$. By Lemma 6.4.8, M is q-nary $(d, r]$-disjunct. Using M as the q-nary matrix in Theorem 6.4.7, we obtain Theorem 6.4.9. □

Some examples were given in [8] to illustrate the effectiveness of Theorem 6.4.9.

$$t(2, 2, n) \le \binom{r}{2} \text{ for any } n \ge 4 \text{ (Corollary 6.4.2)},$$

$$t(2, 2, 16) = t(2, 2, 4^2) \le t(2, 2, 4) \cdot (2 \cdot 2 + 1) \le \binom{4}{2} \cdot 5 = 30,$$

$$t(2, 2, 256) = t(2, 2, 16^2) \le t(2, 2, 16) \cdot 5 \le 30 \cdot 5 = 150,$$
$$t(2, 2, 4096) = t(2, 2, 16^3) \le t(2, 2, 16) \cdot (2^3 + 1) \le 30 \cdot 9 = 270.$$

Another construction is to use the hashing family.

A (t, n, q, m)-*k-perfect hash family* is a $t \times n$ q-nary matrix while for any m of the n columns, there exist at least k rows such that the entries of the m columns are distinct. Note that a (t, n, q, m)-k-perfect family is a q-nary $(d, r; k]$-disjunct matrix for any d, r satisfying $d + r = c$. Wang and Xing [25] proved

Lemma 6.4.10 *There exists an explicit construction for an infinite family of* (t, n, q, m)-*1-perfect hash family with* $t = O(c(m) \log n)$ *for some function of* m.

Using this perfect hash family as a q-nary $(d, r; 1]$-disjoint matrix in Theorem 6.4.4, we obtain

Corollary 6.4.11 *There exists an infinite class of* $t \times n$ q-*nary* $(d, r; z]$-*disjunct matrices with* $t = O(zc(m) \log n)$.

Define $\log^*(1) = 1$ and $\log^*(n) = \log^*(\lceil \log n \rceil) + 1$ for $n > 1$. Stinson, Wei and Zhu also used the separating hash families to prove

Lemma 6.4.12 *There exists an infinite class of* $t \times n$ q-*nary* (d, r)-*disjunct matrices with* $t = O((dr)^{\log^*(n)} \log n)$.

Using this q-nary matrix and z copies of the matrix in Theorem 6.4.1 as inputs to the mechanism in Theorem 6.4.4, we have

Theorem 6.4.13 *There exists a* $t \times n$ $(d, r; z]$-*disjunct matrix with* $t = O(z\binom{n}{z}(dr)^{\log^*(n)} \log n)$.

Stinson and Wei used the vertex z-cover of a hypergraph to construct $(d, r; z]$-disjunct matrices.

For a hypergraph (V, E), (note that this hypergraph has nothing to do with the underlying hypergraph H), VC_z is a *vertex z-cover* if for every edge $e \in E$, there exist at least z vertices $v \in VC_z$ such that $v \in e$. Then VC_1 is simply a vertex-cover. Stinson and Wei interpreted a $(d, r; z]$-disjunct matrix as a vertex k-cover of some hypergraph.

Let $V_{n;\ell,u} = \{X \subseteq [n] \mid \ell \leq |X| \leq u\}$, where $0 < \ell < u < n$. Define a class of *order-interval hypergraph* $H_{n;\ell,u}(V, E)$ with $V = V_{n;\ell,u}$,

$$e_{X,Y} = \{S \mid X \subseteq S \subseteq Y\},$$

and

$$E = \{e_{X,Y} \mid |X| = \ell, |Y| = u\}.$$

Theorem 6.4.14 *There exists a vertex* z-*cover* Z *of* $H_{n;\ell,u}$ *of size* t *if and only if there exists a* $t \times n$ $(\ell, n - u; z]$-*disjunct matrix.*

Proof. Let M denote the incidence matrix where rows are labeled by Z and columns by $[n]$. Consider any $l+n-u$ columns $C_1, ..., C_{l+n-u}$. Define $X = \{C_1, ..., C_l\}$ and $Y = \{C_i \mid i \notin \{l+1, ..., l+n-u\}\}$. Then there exist z rows $R_1, ..., R_z$ such that $X \subseteq R_i \subseteq Y$ for $1 \le i \le z$, or equivalently, R_i intersects $C_1, ..., C_l$ but not $C_{l+1}, ..., C_{l+n+u}$ for $1 \le i \le z$. Hence M is $(\ell, n-u; z]$-disjunct. □

Stinson and Wei derived a lower bound of a minimum z-cover to serve as a lower bound of $t(d, r, n; z]$. Define

$$\tau^z_{n;\ell,u} = \min\{|Z| \mid Z \text{ is a vertex } z\text{-cover of } H_{n;\ell,u}\}.$$

A *fractional vertex d-cover* is a function $g : V \to R^+$ such that for any $e_{X,Y} \in E$,

$$\sum_{v \in e_{X,Y}} g(v) \ge z.$$

Define

$$(\tau^*)^z_{n;\ell,u} = \min\{\sum_{v \in V} g(v) \mid g \text{ is a fractional } z\text{-cover of } H_{n;\ell,u}\}.$$

(The superscript z will be omitted for $z = 1$.)

Stinson and Wei, extending ideas of Engel [9] from $z = 1$ to general z, obtained the following inequalities by choosing g properly:

$$\tau^z_{n,\ell,u} \ge (\tau^*)^z_{n,\ell,u}, \tag{6.4.1}$$

$$(\tau^*)^z_{n,\ell,u} = z \times (\tau^*)_{n,\ell,u}, \tag{6.4.2}$$

$$\tau^z_{n,\ell,u} \ge (\tau^*)^z_{n,\ell,u} \times (\tau^*)_{u-\lambda;\ell-\lambda,u-\lambda}, \tag{6.4.3}$$

$$\tau^z_{n,\ell,u} \ge (\tau^*)_{n,\ell,u} \times (\tau^*)^z_{u-\lambda;\ell-\lambda,u-\lambda},. \tag{6.4.4}$$

From (6.4.1) and (6.4.2), results on $(\tau^*)_{n,\ell,u}$ can be translated to results on $\tau^z_{n,\ell,u}$, and hence on $t(d, r, n; z]$. In particular, using a result on τ^* of Engel, we obtain

Theorem 6.4.15 $t(d, r, n; z] \ge \min\{z\binom{n}{z}/\binom{n-d-1}{m-r} \mid r \le m \le n-d\}$.

Using (6.4.3), we obtain

Theorem 6.4.16

$$t(d, r, n; z] \ge \min_{\substack{r-\lambda_1 \le m_1 \le n-d+\lambda_2 \\ \lambda_1 \le m_2 \le n-d-r+\lambda_1}} \frac{z\binom{n}{m_1}\binom{n-d-r+\lambda_1+\lambda_2}{m_2}}{\binom{n-d-r+\lambda_1+\lambda_2}{m_1-r+\lambda_1}\binom{n-d-r}{m_2-\lambda_1}},$$

where $0 < \lambda_1 < r_1, 0 < \lambda_2 < d$ *and* m_1 *and* m_2 *are integers.*

Using (6.4.4), we obtain

Theorem 6.4.17

$$t(d,r,n;z] \geq \min_{r-\lambda_1 \leq m \leq n-d+\lambda_2} \frac{\binom{n}{m}}{\binom{n-d-r+\lambda_1+\lambda_2}{m-r+\lambda_1}} t(\lambda_2, \lambda_1, n-d-r+\lambda_1+\lambda_2; z],$$

where $0 < \lambda_1 < r$, $0 < \lambda_2 < d$ and m is an integer.

Setting $\lambda_1 = r-1$ and $\lambda_2 = d-1$, it is easily checked that the right hand side attains minimum at $m = n/2$. Thus

Corollary 6.4.18 $t(d,r,n;z] \geq 4(1-\frac{1}{n})t(d-1,r-1,n-2;z]$.

Using this recursion and (2.6.1), we obtain

Theorem 6.4.19

$$\begin{aligned} t(d,r,n;z] \;&\geq\; c4^{r-1}(1-\frac{1}{n})(1-\frac{1}{n-1})\cdots(1-\frac{1}{n-2r+2})[\frac{(d-r+1)^2}{\log(d-r+1)}\log(n-2r) \\ &+(z-1)(d-r+1)], \end{aligned}$$

where c is the same constant as in (2.6.1).

6.5 Random Designs

An attempt to extend the analysis of random designs (Chapter 5) from group testing to graph testing encounters some basic difficulties. Consider a hypergraph H and a fixed but unknown subgraph D. Let $\mathcal{D}$ be the sample space of D, i.e., a member of $\mathcal{D}$ is a set of edges which is a candidate of D. The problem is to identify D using edge tests. In the group testing problem, D is just a set of vertices. Due to symmetry, the analysis for D and D' are the same if $|D| = |D'|$. But in the graph testing problem, even if $|D| = |D'|$, the subgraph structure plays an important part in the analysis. Then one has to deal with the large number of subgraph structures and to take average over all D in $\mathcal{D}$.

To alleviate this problem, we make the following assumptions:
(i) $H = H_r^*$ (H has n vertices),
(ii) $|D| = d$ and every edge is equally likely to be in D.
(iii) only RID is studied.

Note that even under the set of assumptions, the structure of D can vary, i.e., D could be a set of r disjoint edges, or an r-star.

Let M be a testing matrix and $I_X(D,M)$ the event that M contains a row covering X but none of the edges in $D \setminus X$. When X is a negative edge and D is the positive set, $I_X(D,M)$ is the event that M identifies X. In Section 6.6, we will see that even if X is positive, $I_X(D,M)$ is very relevant to the identification of X. Thus $I_X(D,M)$

is the focus of our study in random designs. Let $\bar{I}_X(D,M)$ denote the opposite of $I_X(D,M)$. Then

$$\bar{I}_X(D,M) = \cup_i \bar{I}_X(D,M_i)$$

where M_i is the ith row of M.

Although row i and row j are independent in M, the two events $\bar{I}_X(D,M_i)$ and $\bar{I}_X(D,M_j)$, when D is a variable, are positively correlated since if D is a favorable structure, then it favors both rows i and j. Therefore,

$$E_D Prob(\bar{I}_X(D,M)) \geq \prod_i E_D Prob(\bar{I}_X(D,M_i)), \tag{6.5.1}$$

namely, we cannot compute the left hand side by computing the much simpler right hand side. Note that this problem does not exist in group testing since D is invariant given $|D| = d$.

Suppose M is a $t \times n$ RID(p). Define

$$\Phi(t,n,d,r) = E_X E_{D,|D|=d} Prob(I_X(D,M)) = E_{D,|D|=d} Prob(I_X(D,M)) \tag{6.5.2}$$

since H_r^* is edge-transitive.

From (6.5.1), we have

Lemma 6.5.1 $\Phi(t,n,d,r) \leq 1 - [1 - \Phi(1,n,d,r)]^t$.

We will study $\Phi(1,n,d,r)$ first.

Macula, Rykov and Yakhanin [16] proved a lower bound.

Lemma 6.5.2 $\Phi(1,n,d,r) \geq p^r\{\sum_{k=1}^r [\binom{n-r}{k}\binom{r}{r-k}\binom{n}{r}^{-1}](1-p^k)\}^d$.

Proof. X appears in the row with probability p^r. For each positive edge D_j in D, suppose it overlaps with X in $r-k$ vertices, with probability

$$\binom{n-r}{k}\binom{r}{r-k}\binom{n}{r}^{-1}.$$

Then the probability that at least one of the remaining k vertices in D_j does not appear in the row is $(1-p^k)$. Since there are d positive edges, we multiply this probability d times to obtain a lower bound since the event $I_X(D_j,M)$ and $I_X(D_i,M)$ are positively correlated. □

To compute $\Phi(1,n,t,r)$ exactly, we have to enumerate the numerous structures of D as shown in (6.5.2). Surprisingly, there is a way to bypass the enumeration.

Let D^+ be the random variable representing the distribution of $X_0, X_1, ..., X_d$, namely, a specification of the set of r columns belonging to each X_i. Let w be the random variable representing the number of the 1-entries in the row. Let $I(D^+, w)$ denote the indicator function such that

$$I(D^+, w) = \begin{cases} 1 & \text{if the row covers } X_0 \text{ but none of } X_1, ..., X_d, \\ 0 & \text{otherwise.} \end{cases}$$

Then

$$\Phi(1, n, d, r) = E_D E_w I(D^+, w).$$

Let W be the set of w 1-entries in the row. Then $\binom{W}{r}$ is the set of complexes covered by the row. Hence X must be chosen from $\binom{W}{r}$. Note that instead of enumerating the structures of D, it is now only a matter of counting in how many ways we can choose d complexes (those in D) from outside of the set of $\binom{w}{r}$ complexes.

Macula, Torney and Villenkin [17] gave the formula

$$\Phi(1, n, d, r) = \sum_{w=0}^{n} \binom{n}{w} p^w (1-p)^{n-w} \binom{w}{r} \binom{n}{r}^{-1} [1 - \binom{w}{r} \binom{n}{r}^{-1}]^d. \tag{6.5.3}$$

Assume the row has weight w. Then the r columns of X must be taken from the W columns with 1-entries, which none of the complexes in D can have 1-entries in all its r columns.

From the last term of (6.5.3), it is obvious that the D complexes choose their columns independently (as in Lemma 6.5.1). In particular, some of them can choose the same set of columns which contradicts the fundamental assumption that complexes are distinct. We now modify (6.5.3) to distinct complexes.

Lemma 6.5.3

$$\Phi(1, n, d, r) = \sum_{w=0}^{n} \binom{n}{w} p^w (1-p)^{n-w} \binom{w}{r} \binom{n}{r}^{-1} \left[\binom{\binom{n}{r} - \binom{w}{r}}{d} \Big/ \binom{\binom{n}{r} - 1}{d} \right]. \tag{6.5.4}$$

Torney [23] gave a formula from the viewpoint that for fixed X and D, how to choose the $\binom{w}{r} - 1$ complexes (other than X) present in the row.

Lemma 6.5.4

$$\Phi(1, n, d, r) = \sum_{w=r}^{n} \binom{n-r}{w-r} p^w (1-p)^{n-w} \left[\binom{\binom{n}{r} - d - 1}{\binom{w}{r} - 1} \Big/ \binom{\binom{n}{r} - 1}{\binom{w}{r} - 1} \right]. \tag{6.5.5}$$

Proof. Again, w is the row weight. From the bracket term, each complex in $X \cup D$ is treated as fixed (taking a fixed set of columns), which the $\binom{w}{r} - 1$ complexes present in the row other than X must be chosen outside of $X \cup D$. □

Offhand, this approach seems doubtful since a random choice of $\binom{w}{r}-1$ complexes outside of $X \cup D$ does not guarantee the union to be a w-set. While we do not have a direct proof of Lemma 6.5.3 with insight, we provide a mechanical proof that the two RHSs of (6.5.4) and (6.5.5) are indeed equal by noting:

(i) $\sum_{w=0}^{n}$ in (6.5.4) can be changed to $\sum_{w=r}^{n}$ due to the presence of the term $\binom{w}{r}$.

(ii) $\binom{n}{w}\binom{w}{r}\binom{n}{r}^{-1} = \binom{n-r}{w-r}$.

(iii)

$$\begin{aligned}\frac{\binom{\binom{n}{r}-\binom{w}{r}}{d}}{\binom{\binom{n}{r}-1}{d}} &= \frac{[\binom{n}{r}-\binom{w}{r}]\cdots[\binom{n}{r}-\binom{w}{r}-d+1]}{[\binom{n}{r}-1]\cdots[\binom{n}{r}-d]} \\ &= \frac{[\binom{n}{r}-d-1]\cdots[\binom{n}{r}-\binom{w}{r}-d+1]}{[\binom{n}{r}-1]\cdots[\binom{n}{r}-\binom{w}{r}+1} \\ &= \frac{\binom{\binom{n}{r}-d-1}{\binom{w}{r}-1}}{\binom{\binom{n}{r}-1}{\binom{w}{r}-1}}.\end{aligned}$$

Let w_i denote the weight variable of row i. Then

$$E_{\{w_1,\ldots,w_d\}}E_{D^+}\prod_{i=1}^{t} I(D^+, w_i) \neq \prod_{i=1}^{t} E_{\{w_1,\ldots,w_d\}}E_{D^+}I(D^+, w_i) = 1-[1-\Phi(1,n,d,r)]^t,$$

since $\prod$ is not a linear function.

Macula, Torney and Villenkin gave a bound by using a truncated inclusion-exclusion formula.

Theorem 6.5.5

$$\begin{aligned}\Phi(t,n,d,r) \;\geq\; & t\Phi(1,n,d,r) - \binom{t}{2}\sum_{w_2+w_1+w_1'}\binom{n}{w_2}\binom{n-w_2}{w_1}\binom{n-w_2-w_1}{w_1'}p^{2w_2} \\ & \cdot[p(1-p)]^{w_1+w_1'}(1-p)^{2(n-w_2-w_1-w_1')}\binom{w_2}{r}\binom{n}{r}^{-1} \\ & \cdot[\binom{n}{r}-\binom{w_2+w_1}{r}-\binom{w_2+w_1'}{r}+\binom{w_2}{r}]^d\binom{n}{r}^{-d}.\end{aligned}$$

Proof. $t\Phi(1,n,d,r)$ obviously over estimates $\Phi(t,n,d,r)$ since if k rows satisfy $I_X(D,M)$, then this one case is counted k times. We now explain the second term. Let i and i' be two rows both satisfying $I_X(D,M)$. Suppose w_2 columns intersect both i and i', w_1 (w_1') columns intersect i (i') but not i' (i), and $n-w_2-w_1-w_1'$ columns intersect neither. Then X must be covered by the w_2 columns. For $D_j \in D$, it can neither be covered by the w_2+w_1 columns of row i, nor the w_2+w_1' columns of row i'. But when D_j is covered by the w_2 columns, it is subtracted twice and needs to be added back once. □

6.6 Trivial Two-stage Pooling Designs for Complete r-graphs

A properly constructed random design M could have a high probability of containing a row R_i which covers a positive complex X_j but none of the other positive complexes. However R_i cannot identify X_j unless all other complexes covered in R_i are resolved negative complexes (by appearing in rows with negative outcomes). To increase its chance of happening, we may construct a set of pools obtained by taking intersection of R_i with a set M' of row vectors of the same length. Note that all these intersection rows preserve the property that no positive complex other than X_j can appear. Suppose R_i also covers a negative complex X_0. If M' has a row R' containing X_j but not X_0, then the intersection row R_iR' contains X_j but not X_0, and has a positive outcome. Of course, M' can also have a row R'' containing X_0 but not X_j. Then R_iR'' has a negative outcome. Therefore, by collecting the intersection rows with positive outcomes, the probability that X_j is the only complex appearing in all of them is increased.

As we do not know which row in M is R_i, typically we take intersections of every row in M with every row in M' to obtain a new pooling design M''. The pools in M'' are tested and analyzed to generate a set CP of candidates of positive complexes. Some positive complexes would be missed by CP, and some negative complexes would be wrongly picked up by CP. So

$$P^+ = Prob(\text{a positive complex not in } CP).$$

We eliminate unresolved negative complexes by introducing a second stage which confirms or rejects the candidates by individual testing.

A 2-stage design is evaluated by two criteria representing performance and cost, respectively.

(i) P^+,

(ii) the number of tests = the number of pools in $M'' + |CP|$.

Clearly, these two criteria depend on M, M' and CP. For the 2-stage designs studied in this section M is always an RID with parameter p, while M' can be either probabilistic or deterministic, or even related to M. We introduce two CP which have been studied in the literature.

Define

$$M''(i) = \{\text{row } (i, i') \text{ in } M'' \mid 1 \le i' \le t'\}$$

and

$$U(i) = \{\text{rows in } M''(i) \text{ with positive outcomes}\}.$$

Let $C(i)$ denote the set of columns containing $U(i)$. Macula, Torney and Villenkin [17] first introduced the *complex CP*: $X \in CP$ if $\cap X = U(i) \neq \emptyset$ in $M''(i)$. The reason to exclude the $U(i) = \emptyset$ case is to avoid picking up too many negative clones which are simply not present in R_i.

Theorem 6.6.1 *For a 2-stage design M'' under the complex CP,*

$$P^{+} = 1 - \Phi(t, n, d-1, r) Prob(M' \text{ contains a given complex}). \qquad (6.6.1)$$

Proof. Let R_1 be a row in M which covers X_1 but none of the other positive complexes. Then each row in $U(i)$ does not cover any of the other positive complexes, and hence must cover X_1 to have a positive outcome. Hence $\cap X_1 \supseteq U(i)$ in $M''(i)$. But it is also clear that $\cap X_1 \subseteq U(i)$ since X_1 is positive. Hence $\cap X_1 = U(i)$ and $X_1 \in CP$. The second term in (6.6.1) is added to exclude the case $U(i) = \emptyset$. □

A negative complex X_0 can enter CP if both conditions (i) and (ii) are met:

(i) M has a row R_i covering X_0 and a nonempty set S of positive complexes.

(ii) Every row in M' covering X_0 covers a complex of S and vice versa, i.e., $\cap X_0 = \cup_{X_i \in S}(\cap X_i)$.

The requirement on covering positive complexes is to assure that the rows in $\cap X_0$ all have positive outcomes.

Let $P^{(-)}$ denote the probability $\cap X_0 = U(i)$, i.e., X_0 is misclassified into CP. Then

Theorem 6.6.2 *Under the complex CP,*

$$\begin{aligned} P^{(-)} &= \sum_{w=r}^{n} \binom{n}{w} p^{w}(1-p)^{n-w} \binom{w}{r} \binom{n}{r}^{-1} \\ &\cdot \sum_{|S|=1}^{d} \sum_{s=r}^{w} \binom{w}{s} \left[\binom{\binom{s}{r} - I_{X_0 \in US}}{|S|} \binom{\binom{n}{r} - \binom{s}{r}}{d-|S|} \binom{\binom{n}{r}}{d}^{-1} \right] \cdot Prob(cond\ (ii)), \end{aligned}$$

where $I_{X_0 \in US}$ is the indicator function of $X_0 \in US$.

Proof. Suppose row i in M has weight w and covers X as well as the set S of positive complexes such that $|US| = s$. Then these s columns must be chosen from the w columns constituting the weight. The probability of choosing $|S|$ positive complexes from these s columns and the other $d - |S|$ positive complexes not from these s columns is the [ ] term. □

Corollary 6.6.3 $E(|CP|) = d\Phi(t, n, d-1, r) Prob(M' \text{ contains a given complex}) + [\binom{n}{r} - d]P^{(-)}$.

Theorem 6.6.4 *Suppose M' is the complement (interchanging 0 and 1) of an r-separable matrix. Then $E(|CP|) \leq t$ under the complex CP.*

Proof. Let X and X' be two complexes covered by row i of M. Then the property $\cup X \neq \cup X'$ in an r-separable matrix is translated to $\cap X \neq \cap X'$ in M'. Hence at most one X satisfying $\cap X = U(i)$. □

Theorem 6.6.5 *Suppose M' is RID(p'). Then*

$$Prob(condition\ (ii)) \geq \{\sum_{w=r}^{n} \binom{n}{w} (p')^w (1-p')^{n-w} [\binom{w}{r}\binom{n}{r}^{-1} (1 - \left(\binom{n}{r} - \binom{w}{r}\right)_{|S|}) + (1 - \binom{w}{r}\binom{n}{r}^{-1}) \left(\binom{n}{r} - \binom{w}{r}\right)_{|S|}]\}^t.$$

Proof. Given a row y with weight w, $\binom{w}{r}\binom{n}{r}^{-1}$ is the probability that y covers X_0, $\left(\binom{n}{r}-\binom{w}{r}\right)_{|S|}$ is the probability that y does not cover any complex of S. Then the formula gives the probability of the event E_y that y covers X_0 if and only if it covers a complex of S. It is a lower bound since $\{E_y\}$ are positively correlated over the rows, while the formula treats them as independent. □

We can improve the 2-stage procedure by screening the complexes in CP before individually testing them. A complex X admitted to CP by satisfying $\cap X = U(i)$ can be removed if it appears in any test of $C(j)$ with a negative outcome. Further, suppose M has k rows covering X. Then condition (ii) must be satisfied k times for X not to be removed. Usually, k is not very small and $P^{(-)}$ would then tend to 0, essentially eliminating the need of a second stage.

Macula, Torney and Villenkin [17] proposed to set $M' = M$. Write M'' as M^2. Since $(i,j) = (j,i)$, M^2 contains only one of them labeled by $\{i,j\}$. Rows $\{i,j\}$, $1 \leq i \leq t'$, are also deleted. Thus M^2 has $\binom{t}{2}$ rows. Theorem 6.6.1 is reduced to

Theorem 6.6.6 *For M^2 under the complex CP proposed by Macular and Popyack [25], $P^+ = 1 - \Phi(t, n, d-1, r)$.*

Theorem 6.6.6 was given in [17] under the condition $|\cap X| \geq 2$ in M. But this condition is unnecessary since if $|\cap X| = 1$, then $U(i) = \emptyset$; but $\cap X_1 = \emptyset = U(i)$. So X is still in CP. The case $|\cap X| = 0$ is already counted in the formula for P^+.

For two complexes X_1 and X_2, let $\lambda(t,n,d,r)$ denote the probability that $\cap X_1 = \cap X_2$.

Theorem 6.6.7

$$\lambda(t,n,d,r) = \sum_{k=0}^{r} \binom{r}{k}\binom{n-r}{r-k}\binom{n}{r}^{-1} \{1 - p^k + p^k[p^{2r-2k} + (1-p^{r-k})^2]\}^t.$$

Proof. Let $|X_1 \cap X_2| = k$, then given the r columns of X_1, $\binom{r}{k}\binom{n-r}{r-k}$ is the number of ways to choose X_2. To have $\cap X_1 = \cap X_2$, each row either covers both X_1 and X_2, or covers neither. The former event has probability $p^k p^{2(r-k)}$, while the latter has probability $(1-p^k) + p^k(1-p^{r-k})^2$, combining the probabilities of two mutually exclusive sub-events: either row misses one of the common k columns, or it has them all, but missing one of the non-common column in both X_1 and X_2. □

A formula of $|CP|$ is given in [17] using $\lambda(t, n, d, r)$.

Theorem 6.6.8 $E(|CP|) = t\binom{n}{r}\lambda(t, n, d, r)$.

Proof. For each row i, suppose there exists a positive complex X satisfying $\cap X = U(i)$. Then a complex X' can satisfy $\cap X' = U(i)$ only if $\cap X' = \cap X$ in M', with probability $\lambda(t, n, d, r)$. There are t choices for i, and $\binom{n}{r}$ (or $\binom{n}{r} - 1$) choices of X', hence the formula. □

This formula is not exact (besides the subtraction of 1 from $\binom{n}{r}$) in two aspects:

(i) The probability of the condition $\cap X = U(i)$ is not included. Further, since this probability and $\lambda(t, n, d, r)$ are correlated, they cannot be simply multiplied together.

(ii) The formula does not take into consideration the event that even if $\cap X = U(i)$ is not satisfied for all positive X, we could still have $\cap X_0 = U(i)$ for some negative complex X_0, as described in the paragraph after Theorem 6.4.1.

Next we introduce the other CP proposed by Macula and Popyack. Let $C(i)$ denote the set of columns in M_i'' containing $U(i)$.

The *column CP*. $X \in CP$ if $X = C(i)$ and $|C(i)| = r$.

If $C(i) = C(j)$, then only one of them needs to be tested in stage-2. By construction, $|CP| \leq t$.

Theorem 6.6.9 *For M'' under the column CP,*

$$P^+ = 1 - \Phi(t, n, d-1, r)Prob(\text{given } M, \cap X_0 \not\subseteq C \text{ for any column } C \text{ in } M').$$

Proof. The condition $\cap X_0 \not\subseteq C$ assures that C does not contain $U(i)$, i.e., $C \notin C(i)$. □

Corollary 6.6.10 *Suppose M' is an RID with parameter p'. Then under the column CP,*

$$P^+ = 1 - \Phi(t, n, d-1, r) \sum_{k=1}^{t'} \binom{t'}{k} [(p')^r]^k [1 - (p')^r]^{t'-k} [1 - (p')^k]^{n-r}.$$

Proof. k is the number of rows covering X_0 in M'. None of the $n - r$ columns not in X_0 can contain these r rows. □

Macula and Popyack used a different approach to approximate P^+. They compute the probability $\Phi(1, n, C, r)$ that a given row covers X but does not intersect the column C (call such a C a *success*):

$$\Phi(t, n, C, r) = 1 - [1 - \Phi(1, n, C, r)]^t.$$

Since

$$1 - \Phi(t, n, d-1, r) \neq [1 - \Phi(t, n, C, r)]^{d-1},$$

due to positive correlation between $\Phi(1, n, C, r)$ and $\Phi(1, n, C', r)$, they turned to asymptotic analysis by computing the expected number of successful C to be

$$\beta = (n - r)[1 - (1 - p)(p)^r]^t, \tag{6.6.2}$$

which can be approximated by a Poisson variable with mean β. Then the probability that this number is 0 is $e^{-\beta}$. Thus

Theorem 6.6.11

$$P^+ \sim 1 - e^{-\beta}. \tag{6.6.3}$$

Macula and Popyack actually allowed r_i to vary from complex to complex. Let $r^* = \max r_i$. They modified the column CP by admitting $C(i)$ to CP if $|C(i)| \leq r^*$. For each $C(i) \in CP$, test $C(i) \setminus C$ for each $C \in C(i)$ and confirm $C(i)$ as a positive complex only if $C(i) \setminus C$ tests negative for all $C \in C(i)$. Note that the number of stage-2 tests is inflated to $|CP|(r^* + 1)$.

Macula, Rykov and Yekhanin observed that the complement (exchanging 0 with 1) of a r-disjunct matrix satisfies the requirement that for any $r+1$ columns $C_0, C_1, ..., C_r$, there exists a row intersecting $C_1, ..., C_r$, but not C_0. Let $C_1, ..., C_r$ be the columns in a positive complex X. Then for each column C_0 not in X, M' has a row covering X but not C_0. To reduce the number of rows in M' they proposed to use the complement of an *α-almost r-disjunct matrix*, meaning the probability that a random set X of r columns has probability at least α to have no other column C containing $\cap X$. Then Theorem 6.6.9 is reduced to

Theorem 6.6.12 *Suppose M' is the complement of an α-almost r-disjunct matrix. Then under the column CP,*

$$P^+ = 1 - \Phi(t, n, d-1, r)\alpha.$$

Macula, Rykov and Yekhanin also commented that the q-nary MDS code used in Chapter 3 to construct d-disjunct matrices can be used to construct α-almost r-disjunct matrices with $r > d$ and $\alpha \to 1$.

6.7 Sequential Algorithms for H_r

Chang and Hwang [3] first cast the group testing problem on graphs. They formulated the problem of identifying a unique positive item in a set A and a unique positive item in a set B as a problem of identifying a unique edge in a bipartite graph $G(A, B)$. Throughout this section, we assume G (or H) is a graph (hypergraph) with n vertices and edge set E.

Theorem 6.7.1 *$t(G, 1) = \lceil \log |E| \rceil$ if G is a complete bipartite graph.*

Note that $\lceil \log |E| \rceil$ is the trivial information-theoretic lower bound. Chang and Hwang also conjectured that Theorem 6.7.1 holds even if the bipartite graph is not complete, but contains exactly 2^k edges for some k.

The proof of Theorem 6.7.1 uses group tests (as we pointed out in Section 6.1 that for $d = 1$ group tests can be translated into edge-tests). Aigner [1] brought out the edge-test notion explicitly and conjectured for a general graph G,

$$t(G, 1) \leq \lceil \log |E| \rceil + c.$$

Du and Hwang [6] sharpened the conjecture by setting $c = 1$, which was proved by Damaschke [5].

Theorem 6.7.2 $t(G, 1) \leq \lceil \log |E| \rceil + 1$.

There are many examples of G for which $t(G, 1) = \lceil \log |E| \rceil + 1$.
Triesch [24] extended Theorem 6.7.2 to H_r by proving

Theorem 6.7.3 $t(H_r, 1) \leq \lceil \log |E| \rceil + r - 1$.

Triesch gave the proof using the group-test (as versus edge-test) terminology for easier description. Every vertex in the edge in D is considered a *positive vertex.*

A *vertex-cover* of H is a vertex-set which intersects every edge of H. Choose a vertex-cover $C = \{v_1, ..., v_s\}$ by the following greedy algorithm. Let $H^1 = H$ and v_1 be the vertex with maximum degree in H^1. For $2 \leq i \leq s$, let H^i be the hypergraph obtained from H^{i-1} by deleting $v_1, ..., v_{i-1}$ (and their edges), and let v_i denote the vertex with maximum degree in H^i. Let $d_{H^i}(v_i)$ denote the degree of v_i in H^i. It is easily verified

$$d_{H^1}(v_1) \geq d_{H^2}(v_2) \geq \cdots \geq d_{H^s}(v_s).$$

Define

$$\ell_i = \lceil \log |E| / d_{H^i}(v_i) \rceil.$$

Then

$$\sum_{i=1}^{s} 2^{-\ell_i} \leq \sum_{i=1}^{s} 2^{-\log \frac{|E|}{d_{H^i}(v_i)}} = \sum_{i=1}^{s} \frac{d_{H_i}(v_i)}{|E|} = 1.$$

Now, we give the proof of Theorem 6.7.3.

Proof of Theorem 6.7.3. It is proved by induction on the rank r. For $r = 1$, the problem is simply the classic group testing problem with one positive item and hence $t(H_1, 1) = \lceil \log n \rceil = \lceil \log |E| \rceil$. For $r \geq 2$, consider the vertex cover $C = \{v_1, \cdots, v_s\}$ obtained by the greedy algorithm. Since $\sum_{i=1}^{s} 2^{-\ell_i} \leq 1$, by Kraft's Inequality, there is a binary search tree T with leaves $v_1, v_2, \cdots, v_s$ ordering from left to right such that the length of the path from the root to leaf v_i is at most l_i. Each internal node u of T is associated with a test whether there exists a positive vertex among the leaves under the left son of u. Since C is a vertex cover, there must exist a positive vertex in C. Let v_i denote the positive vertex. Denote by T_a the subtree of T rooted at vertex a. One searches this v_i as follows.

begin
    Initially, set $a :=$ the root of T
    while a is not a leaf
        do begin $b :=$ the left son of a;
            test on $\{v_j \mid j$ is a leaf of $T_b\}$;
            if the outcome is negative
                then $a :=$ the right son of a
                else $a := b$;
        end-while;
    $v_i := a$;
end.

Note that this algorithm can find a positive vertex v_i through at most l_i tests. Moreover, when v_i is found, all $v_1, \cdots, v_{i-1}$ have been found to be good. Therefore, the remaining positive vertices are those adjacent to v_i in H_i. The total number of them is $d_{H_i}(v_i)$. Removing v_i from them results in a hypergraph of rank at most $r-1$. By the induction hypothesis, $\lceil \log d_{H^i}(v_i) \rceil + r - 2$ tests are enough to identify the remaining positive vertices. Therefore, the total number of tests is at most

$$l_i + \lceil \log d_{H^i}(v_i) \rceil + r - 2 \leq \lceil \log |E| \rceil + r - 1.$$

□

As commented in Section 6.1, a group test on $S = \{v_1, ..., v_k\}$ can be converted to an edge test in $V \backslash S$ without affecting the testing tree structure, except a "yes" answer to the group test corresponds to a "no" answer to the edge-test. Johann [12] observed that the edge-test version of the Triesch algorithm can still identify a positive edge even when $d > 1$. We will refer to this modification as the *TJ-procedure*. Note that the group testing version identifies the first positive edge while the edge-test version the last; so that their executions may result in different number of tests. But their worst-case performances are identical even though different paths are travelled.

For H a complete r-hypergraph, the $(H_r^*, 1)$ problem is reduced to the classical group testing with n items including r positive ones.

Corollary 6.7.4 $t(r, n) \leq \lceil \log \binom{n}{r} \rceil + r - 1$.

Corollary 6.7.4 is known in classical group testing (Corollary 2.4.2 in [7]).

For $r = 1$, Theorem 6.7.3 is reduced to Theorem 6.7.1. For $d > 1$, Du and Hwang [7] conjectured

$$t(H : d) \leq d(\lceil \log(|E|/d) \rceil + c)$$

for some constant c. (Note that $d \log(|E|/d)$ is the leading term of the information bound.) Johann [12] proved the conjecture with $c = 7$ when H is a graph G. We improve the constant to 6.

Theorem 6.7.5 $t(G : d) \leq d(\lceil \log(|E|/d) \rceil + 6)$.

The proof of Theorem 6.7.5 is based on the construction of a class of algorithms A_ℓ for integer ℓ which requires at most $d(\lceil \log \ell \rceil + 5 - 1/\ell) + |E|/\ell + 1$ tests. By setting $\ell = \lfloor |E|/d \rfloor$, this quantity is at most $d(\lceil \log(|E|/d) \rceil + 6)$.

The algorithm A_ℓ depends crucially on three ideas. The first is that each positive edge is identified by the TJ-procedure (with $r = 2$) in $\lceil \log |E| \rceil + 1$ tests. The second is a method to avoid repeatedly identifying the same positive edge. The third is that it is possible to find a subset of untested edges with proper size under some general condition. Namely,

Lemma 6.7.6 *Suppose S and S' are two disjoint subsets of V and $1 \leq \ell \leq |E|/2$ such that*

(i) $|E(S') \cup E(S, S')| \geq \ell$,

(ii) $\forall v \in S', |E(v, S)| \leq 2\ell$.

Then there exists a subset $W \subseteq S'$ with

$$\ell \leq |E(W) \cup E(W, S)| \leq 2\ell.$$

Proof. Let $\Gamma(v)$ denote the set of v's neighbors. If there is a vertex $v \in S'$ satisfying

$$\ell \leq |\Gamma(v) \cap S| \leq 2\ell,$$

then set $W = \{v\}$ will do. So we may assume $|\Gamma(v) \cap S| < \ell$ for every $v \in S'$. Let $S' = \{v_1, ..., v_k\}$, and let

$$s = \max\{i \in \{1, ..., k\} \mid E(\{v_1, ..., v_i\}) \cup E(\{v_1, ..., v_i\}, S)| < \ell\}.$$

Because of (i), we have $1 < s < k$. Let $W_1 = \{v_1, ..., v_{s+1}\}$. Then

$$|E(W_1) \cup E(W_1, S)| \geq \ell$$

holds. If

$$|E((W_1) \cup E(W_1, S)| \leq 2\ell,$$

then set $W = W_1$ will do. Otherwise, define sets $W_2, W_3, ..., W_s$ recursively through

$$W_i = W_{i-1} \setminus \{v_{i-1}\}, i = 2, ..., s.$$

Since

$$\begin{aligned} & |E(W_{i-1})| + |E(W_{i-1}, S)| - [|E(W_i)| + |E(W_i, S)|] \\ = & |E(W_{i-1})| - |E(W_i)| + |E(W_{i-1}, S)| - |E(W_i, S)| \\ \leq & |E(\{v_{i-1}, \{v_i, ..., v_s\})| + 1 + |E(\{v_{i-1}\}, S)| \\ \leq & |E(\{v_1, ..., v_s\})| + |E(\{v_1, ..., v_s\}, S))| + 1 \\ \leq & \ell, \end{aligned}$$

one of W_i, $2 \leq i \leq s$, will do. □

The algorithm A_ℓ consists of two steps:

Step 1. A_ℓ partitions the n vertices into a number of sets $V_1, ..., V_h$ such that each V_i contains no positive edge and for each $v \in V_i$, $1 \leq i \leq h$, v has a positive edge with each V_j, $j < i$.

Step 2. For each $v \in V_i$, $2 \leq i \leq h$ and $1 \leq j \leq i-1$, identify all positive edges in $\{v\} \cup V_j$.

At the beginning, set $S = \emptyset$ and $S' = V$. In general, when $V_1, ..., V_k$ are nonempty, set $S = V_1$ and $S' = V \setminus \{V_1, ..., V_k\}$. Call a vertex $v \in S'$ *heavy* if $|\Gamma(v) \cap S| > 2\ell$.

For every heavy vertex $v \in S'$, let S_1 denote a set of 2ℓ neighbors of v in S. Test $v \cup S_1$. If positive, identify a positive edge (v, u) and assign v to the first V_i it has no neighbor. If negative, set $\Gamma(v) = \Gamma(v) \setminus S_1$ and repeat the procedure (even though v may not be heavy anymore) until $\Gamma(v) = \emptyset$ (the last S_1 may contain fewer than 2ℓ vertices).

Next, we consider the case that S' contains no heavy vertex. If $|E(S') \cup E(S, S')| \geq \ell$, use Lemma 6.7.6 to find the subset $W \subseteq S'$; if $|E(S') \cup E(S, S')| < \ell$, set $W = S'$. Test $W \cup S$. If negative, assign W to V_1; otherwise, identify a positive edge (u, v). Assign u and v to satisfy the requirement stated in Step 1. Update S and S', and repeat the procedure until $S' = \emptyset$.

We now describe the "assigning" part in more detail. If the test on W is negative, assign W to V_1; otherwise identify a positive edge (u, v) by the TJ-procedure and assign u, v to some V_i and V_j satisfying the condition in Step 1. We show that we only need to go through the assigning procedure for one of them.

Suppose the TJ-procedure identifies vertex v in the vertex-cover. Among the set of vertices adjacent to v, order them so that those in V_I are at the head of order. Now apply the TJ-procedure to identify the first vertex u such that (v, u) is a positive edge. If $u \in V_1$, then we only need to assign v. If $u \notin V_1$, then V_1 has no positive edge to any vertex in W, in particular, to either u or v, hence v can be assigned to V_1 without any testing.

To assign v, we examine whether $v \cup V_i$, $i = 2, 3, ..., k$ contains a positive edge in that order until we find a V_i such that $v \cup V_i$ contains no positive edge. Then we assign v to V_i. If no such V_i exists, we create a new V_{k+1} to store v. To examine $v \cup V_i$, it suffices to consider $\Gamma_i(v) = \Gamma(v) \cap V_i$. If $|\Gamma_i(v)| \leq \ell$, set $W = v \cup \Gamma_i(v)$. Otherwise, let

W be an arbitrary ℓ-subset of $\Gamma_i(v)$. Test $v \cup \Gamma_i(v)$. If negative, set $\Gamma_i(v) = \Gamma_i(v) \setminus W$ and do the same. If we reach $\Gamma_i(v) = \emptyset$, then assign v to V_i. Otherwise, at some stage $v \cup W$ is positive. Use the TJ-procedure to identify a positive edge (v, w). Examine $v \cup V_{i+1}$.

In identifying the positive edges between v and V_j in Step 2, we first remove all u from the latter if (u, v) has already been identified, either as a positive edge or as a negative edge. Let $\Gamma'(v)$ be the set of neighbors of v in V_j. If $|\Gamma'(v)| \leq \ell$, test $v \cup \Gamma'(v)$. Otherwise, let $W \subseteq \Gamma'(v)$ be of size ℓ. Test $v \cup W$. If negative, set $\Gamma'(v) = \Gamma'(v) \setminus W$. If positive, identify a positive edge (v, u) and set $\Gamma'(v) = \Gamma'(v) \setminus \{u\}$. Repeat the procedure with update $\Gamma'(v)$.

We analyze the number of tests required in Step 1. We first count the number of tests consumed in identifying positive and negative edges. Call a test with negative outcome a *bad test* if it contains fewer than ℓ edges. In Step 1 each positive edge in an m-set is identified by at most $\lceil \log m \rceil + 2$ tests, one more than the Triesch's result due to the additional test on $W \cup S$. Each time a positive vertex is assigned, the last test could be bad (all other tests, even negative, are counted in the $\lceil \log m \rceil + 2$ tests in identifying a positive edge). Further, the test on the last S' can be bad. Note that for a heavy vertex, although the test on the last S_1 can be bad, this test must be proceeded by a test on 2ℓ negative edges. So these two tests average out to at least ℓ edges, i.e., we need not count the bad test.

Let d_1 denote the number of positive edges identified and n_1 the number of nonbad tests taken in Step 1. If a positive edge is identified in testing $E(W) \cup E(W \cup V_1)$, then it takes one test for the initial test and at most $\lceil \log 2\ell \rceil + 1$ tests in using the TJ procedure (fewer tests for a heavy item). Further, there is at most one bad test among all tests of this type. If a positive edge is identified in assigning a vertex to some V_i, then it takes $\lceil \log \ell \rceil + 1$ tests in using the TJ procedure and at most one bad test can occur among the negative tests. Therefore the total number of tests in Step 1 is at most

$$d_1(\lceil \log 2\ell \rceil + 2) + n_1 + d_1 + 1.$$

In Step 2, there are at most d_1 vertices not in V_1 since each of them implies a distinct positive edge. In checking the positive edges of $v \in V_i$, the test on $v \cup V_j$ may contain one bad test. But again, each bad test can be assigned to the identification of a positive edge. Let d_2 and n_2 denote the counterparts of d_1 and n_1 in Step 2. Then the number of tests in Step 2 is at most

$$d_2(\lceil \log \ell \rceil + 2) + n_2 + d_1.$$

Adding up, the total number of tests is at most

$$(d_1 + d_2)(\lceil \log \ell \rceil + 5) + n_1 + n_2 + 1.$$

By noting

$$d_1 + d_2 = d$$

$$(n_1 + n_2)\ell \leq |E| - d,$$

we obtain an upper bound

$$d(\lceil \log \ell \rceil + 5) + \frac{|E| - d}{\ell} + 1 = d(\lceil \log \ell \rceil + 5 - \frac{1}{\ell}) + \frac{|E|}{\ell} + 1.$$

The proof of Theorem 6.7.5 is complete.

Recently, Hwang [11] gave a competitive algorithm for graphs with d unknown. Chen and Hwang [4] extended to hypergraphs. They followed Johann's approach in general, but had to resolve some problems unique to hypergraphs.

In Johann's algorithm, each positive edge is broken into two subsets. Should a rank-r edge be broken into two subsets, r subsets or something in between? Later, when searching for positive edges between two subsets, Johann simply takes one vertex u from one subset and then remove $\{(v \mid (u, v)$ is an identified positive edge$\}$ from the other subset to avoid the identification of a positive edge already identified. For the hypergraph case, if an r-edge is broken into r subsets, then we need to mix vertices from more than two subsets; if not, then we need to take more than one vertex from a subset. How do we avoid the identified positive edges? Finally, there is the difficulty in analysis which, to a large degree, depends on the choice of breaking a positive edge.

To avoid being overly complicated, Chen and Hwang chose the simplest setting of partitioning each positive edge into two subsets V_0 and V_1. At the beginning $V_0 = V$. Test V_0, if positive, use the TJ-procedure to identify a positive edge e. Assign an arbitrary vertex v of e to V_1 and set $V_0 = V_0 \setminus \{v\}$. Test V_0 again. Do this until the testing outcome on V_0 is negative. There are still unidentified positive edges e with at least one vertex in V_1. These edges are identified essentially through an iteration process. Let K be a nonempty subset of V_1. We identify all positive edges contained in $K \cup V_1$. But since all positive edges of the type $K' \cup V_1'$, where $K' \subset K$ and $V_1' \subseteq V_1$ were identified earlier when K was set to be K', we only identify positive edges of the type $K \cup V_1'$, where $V_1' = V_1$. This is done by solving a subproblem where $e' \subseteq V_1$ is a positive edge in the subproblem if and only if $K \cup e'$ is a positive edge in the original problem.

To avoid the identified positive edges, let V' be the vertex-set of the subproblem and $C \in V'$ an identified vertex-cover of the positive edges. Then C is moved from V' to V_1' before any testing. Thus in testing V', we will never encounter a positive edge.

Note that the subproblem is the same as the original problem except

(i) r is changed to $r - |K|$,

(ii) K and C are attached to the subproblem.

Thus we will enlarge the original problem to allow K and C (both equal to $\emptyset$ in the original problem) so that the problem can be solved recursively. Note that in the subproblem when the maximum rank is 1 and K is given. Then an identified positive edge can be avoided by removing $\{v \in V \mid K \cup V$ is an identified positive edge$\}$.

We first give an algorithm for r-hypergraphs.

Let K_i denote the subset imposed on $CH(i)$, i.e., K_i is a part of every test in $CH(i)$. If the original problem is defined on an r-hypergraph, then $|K_i| = r - i$. The vertices V_i of $CH(i)$ are divided into V_{i0} and V_{i1}. Define $E(K_i) = \{e \subseteq V_i \mid e \cup K_i \in E\}$ and $D(K_i) = \{e \subseteq V_i \mid e \cup K_i \in D\}$. Finally, let I denote the set of currently identified positive edges (in the original problem). Define $I(K_i) = \{e \subseteq V_i \mid e \cup K_i \in I\}$. For $CH(r)$, $V_{i0} = V$ and $V_{i1} = K_r = I = \emptyset$. Thus $E(K_r) = E$, $D(K_r) = D$ and $I(K_r) = \emptyset$.

We first define $CH(1)$ and then give a recursive algorithm $CH(i)$.

Algorithm $CH(1)$

Input: K_1, V_{10}, V_{11}, I. Attach K_1 to every test.

Step 1. Test V_{10}. If positive, use the halving procedure, which we will treat as a special case of the TJ-procedure, to identify a positive vertex u. Set $I := I \cup \{e\}$, $V_{10} := V_{10} \setminus \{u\}$ and go back to the beginning of Step 1. If negative, go to Step 2.

Step 2. For every vertex $v \in V_{11}$, test $K_1 \cup \{v\}$. If positive, set $I := I \cup \{K_1 \cup \{v\}\}$. If negative, stop.

Algorithm $CH(i)$ $(i \geq 2)$

Input: E, D, K_i, V_{i0}, V_{i1}, I. Attach K_i to every test.

Partition Stage:

Step 1. Test V_{i0}. If positive, use the TJ-procedure to identify a positive edge $e = \{v_1, v_2, ..., v_i\} \subseteq V_{i0}$. Add the vertex v_1 to V_{i1}. Set $V_{i0} := V_{i0} \setminus \{v_1\}$ and $I := I \cup \{K_i \cup \{v\}\}$. If $|V_{i0}| \geq i$, go back to the beginning of Step 1.

Step 2. If V_{i1} is nonempty, go to the search stage.

Search Stage:

Step 1. Set $k = 1$.

Step 2. Let K be a k-subset of V_1. Set $K_{i-k} = K_i \cup K$. Construct a vertex cover C on $I(K_{i-k})$. Call subroutine $CH(i-k)$ with $V_{i-k,0} = V_{i0}$, $V_{i-k,1} = C$, K_{i-k} and I. Do this for all k-subsets K. Set $k := k + 1$. If $k < i$, go back to the beginning of Step 2.

Step 3. Test all i-subsets S (except those such that $S \cup K_i \in I$) of V_{i1}. If positive, set $I := I \cup \{S \cup K_i\}$.

Step 4. Stop.

We will refer to tests in Step 3 as *direct hits.*

By inspecting the algorithm, we note that each positive edge is either identified by the TJ-procedure during a partition stage, or by a direct hit during a search stage. Thus every positive edge is identified in $d(\log|E|+r-1)$ tests. $CH(r)$ does not attempt to optimize the test size as in Johann's algorithm (and pay a price of increasing the leading term from $d\log(|E|/d)$ to $d\log|E|$, but still manage to control the number of negative tests. Since the parameter d is used in the algorithm only in determining the optimal size, $CH(r)$ assumes no knowledge of d and is thus a competitive algorithm.

Theorem 6.7.7 *Let E be an arbitrary r-hypergraph which contains d positive edges, where d is not necessarily known. Then the algorithm $CH(i)$ identifies all positive edges of E with at most $d\lceil\log_2|E|\rceil+(i-1)^{\lfloor i/2\rfloor}d^i+o(d^i)$ tests.*

Proof. Clearly, all edges identified as positive by the algorithm are through either the TJ-procedure or direct hits, both are error-free. Thus it suffices to prove that a positive edge is always identified.

Suppose a positive edge with vertex set X is not identified at the partition stage of $CH(i)$. Then a nonempty subset $X'\subseteq X$ must lie in V_{i1}. At the search stage, the selection of K runs through all k-subsets of V_{i1} for $1\le k\le i$. One such selection is $K=X'$. Suppose $|X'|=k$. Then the problem is reduced to the subroutine $CH(i-k)$ with K imposed. By induction on i, the induced positive edge $X\setminus K$ can be identified in the subroutine, which implies $(X\setminus K)\cup K=X$ is a positive edge.

It remains to count the number of tests $CH(i)$ uses. Note that TJ-procedure uses at most $\lceil\log_2|E|\rceil+i$ tests. Since a positive edge is identified by either the TJ-procedure or a direct hits, the number of tests consumed in identifying one positive edge is bounded by $\lceil\log_2|E|\rceil+i$. This bounded number of tests includes the possible positive test initiating the identification process, and all negative tests occurred during the process of identifying the positive edge. Thus, the number of tests identifying d positive edges is at most $d(\lceil\log_2|E|\rceil+i)$.

Further, it suffices to count the number $N(i)$ of negative tests occurred elsewhere in $CH(i)$. There are three sources for negative tests in $N(i)$: one negative test from the partition stage, those from the subroutines and direct hits.

Denote D_k as the set of all positive edges in $K\cup V(E_0(K))$. Let $d_K=|D_K|$. Note that for $K\neq K'$, D_K and $D_{K'}$ may overlap in positive edges containing some vertices in $K\cap K'$ and other vertices in $V(E_0(K))$. Hence we can only bound d_K by d. However for $|K|=1$, D_K and $D_{K'}$ are disjoint; hence $\sum_{K:|K|=1}d_K$ is bounded by d. We count the number $N(i)$ of negative tests in $CH(i)$ by induction on i.

For $i=1$, $N(i)$ is easily verified to be at most 1. Since each positive vertex can be identified by the halving procedure in $\lceil\log_2|E|\rceil$ tests, Theorem 6.7.7 holds for $i=1$.

We prove the general $i\ge 2$ case by induction.

$$N(i)\ \le\ 1+\sum_{k=1}^{i-2}\sum_{K\subseteq V_{i1}:|K|=k}N(i-k)+\sum_{K\subseteq V_{i1}:|K|=i-1}N(1)+\binom{|V_{i1}|}{i}$$

$$
\begin{aligned}
&\le \sum_{k=1}^{i-2} \sum_{K\subseteq V_{i1}:|K|=k} ((i-k-1)^{\lfloor\frac{i-k}{2}\rfloor} d_K^{i-k} + o(d_K^{i-k})) + \sum_{K\subseteq V_{i1}:|K|=i-1} (1+d_k) + d^i \\
&\le \sum_{K\subseteq V_{i1}:|K|=1} (i-2)^{\lfloor\frac{i-1}{2}\rfloor} d_K^{i-1} + \sum_{k=2}^{i-2} \sum_{K\subseteq V_{i1}:|K|=k} ((i-k-1)^{\lfloor\frac{i-k}{2}\rfloor} d_K^{i-k} \\
&\quad + d^{i-1}(1+d) + d^i + o(d^i) \\
&\le (i-2)^{\lfloor\frac{i-1}{2}\rfloor} \sum_{K\subseteq V_{i1}:|K|=1} d_K^{i-1} + (i-3)^{\lfloor\frac{i}{2}\rfloor-1} \sum_{k=2}^{i-2} \binom{d}{k} d^{i-k} + 2d^i + 2d^i + o(d^i) \\
&\le (i-3)^{\lfloor\frac{i}{2}\rfloor-1}(i-3)d^i + 2d^i + o(d^i) \\
&= ((i-3)^{\lfloor\frac{i}{2}\rfloor} + 2)d^i + o(d^i) \\
&\le (i-3)^{\lfloor\frac{i}{2}\rfloor} d^i + o(d^i).
\end{aligned}
$$

Thus, $N(i) \le (i-1)^{\lfloor\frac{i}{2}\rfloor} d^i + o(d^i)$ holds for general i.

Let $T(i)$ denote the total number of tests required by $CH(i)$. Since $T(i) \le d(\lceil\log_2 |E|\rceil + i) + N(i)$ and $N(i) \le (i-1)^{\lfloor\frac{i}{2}\rfloor} d^i + o(d^i)$, we have $T(i) \le d(\lceil\log_2 |E|\rceil + i) + (i-1)^{\lfloor\frac{i}{2}\rfloor} d^i + o(d^i)$ for $i \ge 2$. Therefore, algorithm $CH(i)$ needs at most $d(\lceil\log_2 |E|\rceil + i) + (i-1)^{\lfloor\frac{i}{2}\rfloor} d^i + o(d^i)$ tests to identify all d positive edges in E. □

We now extend the algorithm to general H. Let E be a hypergraph of rank i, i.e., $|e| \le i$ for all edges $e \in E$. To identify the set $D \subset E$ in a hypergraph, we follow the general approach in algorithm $CH(i)$ for i-hypergraph with a slight modification. Let $CH^*(i)$ denote the algorithm for hypergraph of rank i.

The search stage in $CH^*(i)$ will be slightly different from $CH(i)$. When we choose a k-subset K of V_{i1} before constructing a vertex cover and then calling $CH^*(i-k)$, we should test K itself. If the outcome is positive, then $K \in D$. By our assumption, there is no other positive edge containing K, so we do not need to call $CH^*(i-k)$ further. If the outcome is negative, call $CH^*(i-k)$ to identify all induced positive edges in $E_0(K)$.

Algorithm $CH^*(1)$ is the same as $CH(1)$. Now, we give the algorithm $CH^*(i)$ recursively.

Algorithm $CH^*(i)$ $(i \ge 2)$

input: E, K_i, V_{i0}, V_{i1}, I (if $CH^*(i)$ is not a subroutine, then $V_{i0} := V(E)$ and $V_{i1} := K_i := I := \emptyset$).

Partition Stage

Step 1. Test V_{i0}. If positive, use the TJ-procedure to identify a positive edge $e = \{v_1, v_2, ..., v_s\} \subseteq V_{i0}$ $(s \le i)$. Add the vertex v_1 to V_{i0}. Set

$V_{i0} := V_0 \setminus \{v_1\}$ and $I := I \cup \{\{e\} \cup K_i\}$. If $V_{i0} \neq \emptyset$, go back to the beginning of Step 1.

Step 2. If $V_{i1} \neq \emptyset$, go to Search Stage.

Search Stage:

Step 1. Set $k := 1$.

Step 2. Choose a k-subset K of V_{i1}, where $G(K_i \cup K)$ does not contain any identical positive edge in I. Set $K_{i-k} := K_i \cup K$. Test K_{i-k}. If positive, let $I := I \cup \{K_{i-k}\}$. Else, construct a vertex cover C ($C \cap K_{i-k} = \emptyset$) on $I(K_{i-k} \cup V(E_0(K_{i-k})))$. Call subroutine $CH^*(i-k)$ with $E := E_0(K_{i-k})$, $V_i := V(E_0(K_{i-k}))$, $V_{i0} := V(E_0(K_{i-k})) \setminus C$ and $V_{i1} = C$. If for some $v \in V_{i0}$, $K_{i-k} \cup \{v\} \in I(K_{i-k} \cup V(E_0(K_{i-k})))$ (possibly only for $k = i-1$), delete v from V_{i0}. Attach K_{i-k} to any test in $CH^*(i-k)$. Do this for all k-subsets K. Set $k := k+1$. If $k < i$, go back to the beginning of Step 2.

Step 3. Test all i-subsets S (except those such that $S \cup K_i \in I$) of V_{i1}. If positive, set $I := I \cup \{S \cup K_i\}$.

Step 4. Stop.

Theorem 6.7.8 *Let E be a hypergraph of rank r with d positive edges, where d is not necessarily known. Then the algorithm $CH^*(r)$ identifies all positive edges in E with at most $d\lceil \log_2 |E| \rceil + (r-1)^{\lfloor \frac{r}{2} \rfloor} d^r + o(d^r)$ tests.*

Proof. Similar to the proof of Theorem 6.7.7, we can show that $CH^*(r)$ identifies all positive edges of the hypergraph E.

To count the number of tests $CH^*(r)$ uses, let $N*(r)$ and $T*(r)$ be the counterparts of $N(r)$ and $T(r)$ in $CH^*(r)$. The analysis of the test number of $CH^*(r)$ is also similar to that of $CH(r)$. The only difference is that the subroutine of $CH^*(r)$ should need $N*(r-k)+1$ tests instead of $N*(r-k)$ tests in $CH(r)$. But it does not change the result; so $N*(r) \leq (r-1)^{\lfloor \frac{r}{2} \rfloor} d^r + o(d^r)$. Consequently, $T*(r) \leq d\lceil \log_2 |E| \rceil + (r-1)^{\lfloor \frac{r}{2} \rfloor} d^r + o(d^r)$ for $r \geq 2$. Therefore, algorithm $CH^*(r)$ needs at most $d\lceil \log_2 |E| \rceil + (r-1)^{\lfloor \frac{r}{2} \rfloor} d^r + o(d^r)$ tests to identify all d positive edges of E. □

References

[1] M. Aigner, Search problems on graphs, *Disc. Appl. Math.*, 14 (1986) 215-230.

[2] H. B. Chen, D.-Z. Du and F. K. Hwang, An unexpected meeting of four seemily unrelated problems: graph testing, DNA complex secreening, superimposed codes and secure key distribution, preprint, 2005.

[3] G. J. Chang and F. K. Hwang, A group testing problem, *SIAM J. Alg. Disc. Methods*, 1 (1980) 21-24.

[4] T. Chen and F. K. Hwang, A competitive algorithm in searching for many edges in a hypergraph, 2003, preprint.

[5] P. Damaschke, A tight upper bound for group testing in graphs, *Disc. Appl. Math.*, 48 (1994) 101-109.

[6] D.-Z. Du and F. K. Hwang, *Combinatorial Group Testing and Its Applications*, World Scientific, Singapore, 1993.

[7] D.-Z. Du and F. K. Hwang, *Combinatorial Group Testing and Its Applications*, 2nd edition, World Scientific, Singapore, 2000.

[8] A. Dyachkov, P. Villenkin, A. Macula and D. Torney, On families of subsets where no intersection of ℓ-subsets is covered by the union of s others, *J. Combin. Thy. (Series A)*, 99 (2002) 195-218.

[9] K. Engel, Interval packing and covering in the boolean lattice, *Combin. Probab. Compul.*, 5 (1996) 373-384.

[10] H. Gao, F.K. Hwang, M. Thai, W. Wu and T. Znati, Construction of disjunct matrices for group testing in the complex model, manuscript.

[11] F. K. Hwang, A competitive algorithm to find all defective edges in a graph, *Disc. Appl. Math.*, 148 (2005) 273-277.

[12] P. Johann, A group testing problem for graphs with several defective edges, *Disc. Appl. Math.*, 117 (2002) 99-108.

[13] H. K. Kim and V. Lebedev, On optimal superimposed codes, *J. Combin. Design*, 12 (2004) 79-91.

[14] Y. Li, M. Thai, Z. Liu and W. Wu, Protein-to-protein interactions and group testing in bipartite graphs, *International Journal of Bioinformatics and Applications*, 1 (2005) 414-419.

[15] A. J. Macula and L. J. Popyack, A group testing method for finding patterns in data, *Disc. Appl. Math.*, 144 (2004) 149-157.

[16] A. J. Macula, V. V. Rykov and S. Yekhanin, Trivial two-stage group testing for complexes using almost disjunct matrices, *Disc. Appl. Math.*

[17] A. J. Macula, D. C. Torney, and P.A. Villenkin, Two-stage group testing for complexes in the presence of errors, *DIMACS Series in Disc. Math. and Theor. Comput. Sci.*, 55 (2000) 145-157.

[18] C. J. Mitchell and F. C. Piper, Key storage in secure networks, *Disc. Appl. Math.*, 21 (1988) 215-228.

[19] Y. L. Sagalovich, On separating systems, *Problemy Peredachi Informatsii*, 30 (1994) 14-35 (in Russian).

[20] D. R. Stinson, On some methods for unconditionally secure key distribution and broadcast encryption, *Designs, Codes, Crypto.*, 12(1997) 215-343.

[21] D. R. Stinson and R. Wei, Generalized cover-free families, *Disc. Math.*, 27 (2004) 463-477.

[22] D. R. Stinson, R. Wei and L. Chu, Some new bounds for cover-free families, *J. Combin. Thy. Series A*, 90 (2000) 224-234.

[23] D. C. Torney, Sets pooling designs, *Ann. Combin.*, 3(1999) 95-101.

[24] E. Triesch, A group testing problem for hypergraphs of bounded rank, *Disc. Appl. Math.*, 66 (1996) 185-188.

[25] H. Wang and C. Xing, Explicit construction of perfect hash families from algebraic curves over finite fields, *J. Combin. Thy. Ser. A*, 93 (2001) 112-124.

7
Contig Sequencing

In this chapter, we continue the study on the edge-testing model in searching a hidden subgraph except that we assume that we know the topology of the hidden subgraph. Most of the current results focus on two cases. In the first case, the hidden subgraph is a Hamiltonian cycle and in the second case, it is a matching, while the underlying hypergraph is a complete graph. They are both motivated by the contig sequencing problem.

7.1 Introduction

In Section 1.3, we have stated that a multiplex PCR test is supposed to yield the number of pairs of adjacent contigs, but the number is not completely reliable. Traditionally, there are two ways to handle this information. The first is to take this information at its face value. The corresponding model is called the *quantitative multi-vertex model* where a test reveals the number of adjacent pairs it contains. The second way is to ignore this quantitative information. The corresponding model is called *multi-vertex model* where a test simply reveals whether it contains an adjacent pair. A compromise between these two models is a third model, called *incremental model*, where the quantitative information is allowed only in the most restrictive way. Namely, the difference between testing a subset S and a subset $S \cup \{v\}$, where v is a vertex, is distinguishable if $S \cup \{v\}$ contains more adjacent pairs than S. Note that if the same error occurs in testing S and $S \cup \{v\}$, then both carry wrong quantitative information, but the incremental information is intact.

Grebinski and Kucherov [7] cast the contig sequencing problem as an edge-testing problem in graph by treating the contigs as vertices, G being the complete graph and D being either a Hamiltonian path or a Hamiltonian cycle (when the target sequence is circular). Since the path case and the cycle case have no fundamental difference, we will consider the cycle case only, to take advantage of its symmetry. Beigel *et al.* [3] proposed a different graph model where primers are vertices, G is the complete graph minus a 1-factor (the two primers of a contig constitute an edge in the 1-factor), and D is a 1-factor (a complete matching). This formulation has the advantage that each vertex is of degree 1. Hence, once an edge is identified, we can remove its two vertices without worrying another edge being accidentally removed, a major concern in edge-testing.

Theoretically, we can construct a pooling design using the $(d, 2]$-disjunct matrix introduced in Section 6.2, where d is either $n/2$ (contig-vertices) or n (primer-vertices). However, the $(d, r]$-disjunct matrices are efficient only when d is relatively small compared to n. Here we have $d \geq n/2$. Any algorithm given in the last chapter would require a number of tests larger than quadratic in the number of contigs, which is the order of testing all pairs. The inefficiency is due to the fact that the (d, r)-disjunct matrix depends on the size of D but ignores its structure. We will consider algorithms which take advantage of the unique structure of D, i.e., a Hamiltonian cycle with contigs as vertices, or a 1-factor with primers as vertices. In this chapter we assume that the primers are the vertices unless specified otherwise.

7.2 Some Probability Analysis of a k-subset

We first offer some probability analysis on a random k-set.

Consider a complete graph K_v with v vertices and e positive edges. Let $E(v, e, k)$ denote the expected number of positive edges in a random k-clique.

Lemma 7.2.1 $E(v, e, k) = e\binom{k}{2} / \binom{v}{2}$.

Proof. The probability that a positive edge is in the k-clique is $\binom{k}{2} / \binom{v}{2}$. □

Corollary 7.2.2 *Consider v/k k-cliques with no overlapping of edges. Then the expected number # of positive edges contained in these k-cliques is $e(k-1)/(v-1)$. For $e \leq v/2$ and $k < v$, this number $\# < k/2$.*

Proof.

$$\begin{aligned} \# &= (v/k)e\binom{k}{2} \Big/ \binom{v}{2} \\ &= e(k-1)/(v-1). \end{aligned}$$

Further, for $e \leq v/2$ and $k < v$,

$$e(k-1)/(v-1) \leq (v/2)(k-1)/(v-1) < k/2.$$

□

Call a k-clique positive if it contains a positive edge.

Corollary 7.2.3 *The expected number of positive k-cliques is smaller than $k/2$ for $e \leq v/2$ and $k < v$.*

The special case of Corollary 7.2.2 when the k-cliques constituting a partition of the v vertices and the $e = v/2$ positive edges constituting a 1-factor of K_v, was given by Bergel *et al.* Note that in Lemma 7.2.1 and Corollary 7.2.2, the positive edges are not assumed to be vertex-disjoint.

Suppose the e positive edges are disjoint. Then Hwang and Lin [9] gave the pdf $P(i \mid v, e^*, k)$ of i positive edges in a k-clique, where e^* denotes e disjoint positive edges.

Lemma 7.2.4

$$P(i \mid v, e^*, k) = \binom{e}{i}\binom{v}{k}^{-1} \sum_{j=0}^{v-2e} \binom{v-2e}{j}\binom{e-i}{k-2i-j} 2^{k-2i-j}.$$

Proof. There are $\binom{v}{k}$ k-cliques. A clique containing exactly i positive edges must contain both vertices from i positive edges, and a number j, say, from the $v - 2i$ vertices, not incident to a positive edge. Finally, each of the remaining $k - 2i - j$ vertices in the k-clique must be chosen from a distinct edge of the remaining $e - i$ positive edges, with 2^{k-2i-j} such choices. □

Corollary 7.2.5

$$P(i \mid v, (v/2)^*, k) = \binom{v/2}{i}\binom{v}{k}^{-1}\binom{v/2-i}{k-2i} 2^{k-2i}.$$

Lemma 7.2.6 *$P(i \mid v, e^*, k)/P(i-1 \mid v, e^*, k) \leq 1/2i$ if $k^2 - 2e \leq (4i + 2j - 5)k - (2i + j - 2)^2 + j + 2$ for all $j = 0, 1, ..., v - 2e$, and $i \leq \lceil e/2 \rceil$.*

Proof. Express $P(i \mid v, e^*, k)$ and $P(i-1 \mid v, e^*, k)$ as in Lemma 7.2.3 and compare their jth term, $j = 0, 1, ..., v - 2e$.

$$\begin{aligned}
\frac{P(i \mid v, e^*, k)}{P(i-1 \mid v, e^*, k)} &\leq \max_{0 \leq j \leq v-2e} \frac{\binom{e}{i}\binom{e-i}{k-2i-j}}{4\binom{e}{i-1}\binom{e-i+1}{k-2i+2-j}} \\
&= \max_{0 \leq j \leq v-2e} \frac{(k-2i+2-j)(k-2i+1-j)}{4i(e-k+i-j)} \\
&= \max_{0 \leq j \leq v-2e} \frac{k^2 - (4i+2j-3)k + (2i+j-2)(2i+j-1)}{2i - 2(e-k+i+j)} \\
&\leq \frac{1}{2i},
\end{aligned}$$

if the condition in Lemma 7.2.6 is satisfied. □

Denote $P(i \mid v, (v/2)^*, k)$ by $P(i, v, k)$.

Corollary 7.2.7 *$P(i, v, k)/P(i-1, v, k) \leq 1/2i$ if $k^2 - v \leq (4i - 5)k - (2i - 2)^2 + 2$.*

Next we relate the above probability analysis to some Steiner designs. The Steiner design was used in Chapter 3 to construct the d-disjunct matrix, using its property that two blocks intersect in at most one object. Here we use another of its properties that a $(v, k, 1)$-design (Steiner) represents a decomposition of the edge of K_v into edge-disjoint k-clique or blocks. It is easily verified that each object appears in $r = (v-1)/(k-1)$ blocks and there are $b = \binom{v}{2}/\binom{k}{2}$ blocks. A $(v, k, 1)$-design is *resolvable* if the b blocks can be partitioned into r classes each containing v/k blocks which form a partition of the v vertices. It is well known that for p a prime, the $(p^2, p, 1)$-design, known as the *affine plane*, and the $(p^2+p+1, p+1, 1)$-design, known as the *projective plane*, always exist. Tettelin *et al.* [12] were the first to propose using the incidence matrices of such planes to construct pooling designs for contig sequencing.

A Steiner design offers the following major advantages:

(i) If we identify the positive edges in each block, then we identify all positive edges. Therefore the original problem on v vertices is reduced to subproblems each on k vertices, a much smaller scale. There is no waste in this reduction since each edge appears only in one block.

(ii) The work on the b blocks can be performed in parallel. This implies that there is no increase of time in solving the larger problem than the smaller problems.

Suppose both the affine planes and the projective planes are used. Let p be the smallest prime such that either $p^2 \geq n$ or $p^2 + p + 1 \geq n$. Use the affine plane in the former case, and the projective plane in the latter case by adding enough dummy vertices.

Lemma 7.2.8 *If the affine plane with p^2 points is used, then $n \geq p^2 - p + 2$. If the projective plane with $p^2 + p + 1$ points is used, then $n \geq p^2 + 1$.*

Proof. If $n \leq p^2 - p + 1$ in the former case, then we could select the $((p-1)^2 + (p-1) + 1, (p-1) + 1, 1)$ projective plane. If $n \leq p^2$ in the latter case, then we could select the $(p^2, p, 1)$ affine plane. □

Suppose an affine plane $(k^2, k, 1)$ is selected. By Lemma 7.2.8,

$$\begin{aligned} & [(4i-5)k - (2i-2)^2 + 2] - (k^2 - n) \\ \geq\ & [(4i-5)k - (2i-2)^2 + 2] - (k - 2) \\ \geq\ & (4i-8)k - (2i-2)^2 + 4 \\ =\ & 4(i-2)(k-i). \end{aligned}$$

Corollary 7.2.7 can now be improved to

Lemma 7.2.9 *When an affine plane is preferred over a projective plane, then*

$$\frac{P(i, n, p)}{P(i-1, n, p)} \leq \frac{1}{2i} \text{ for } i \geq 2.$$

Suppose a projective plane is selected. Since $k = p + 1$, by Lemma 7.2.8,

$$k^2 - n = (p+1)^2 - n \le (p+1)^2 - (p^2+1) = 2p = 2(k-1).$$

Hence

$$\begin{aligned} & [(4i-5)k - (2i-2)^2 + 2] - (k^2 - n) \\ \ge\ & (4i-5)k - (2i-2)^2 + 2 - 2(k-1) \\ \ge\ & (4i-7)k - (2i-2)^2 + 4 \\ \ge\ & (4i-8)k - (4i^2 - 8i) \\ =\ & 4(i-2)(k-i). \end{aligned}$$

Therefore, we have

Lemma 7.2.10 *When a projective plane is preferred over an affine plane, then*

$$\frac{P(i,n,p+1)}{P(i-1,n,p+1)} \le \frac{1}{2i} \text{ for } i \ge 2.$$

Alon *et al.* [1] computed the probability $\Phi(n, n/2, k)$ that a k-set, known to contain a fixed negative edge e, does not contain any of the $n/2$ disjoint positive edges.

Lemma 7.2.11

$$\Phi(n, n/2, k) = \frac{(n-4)(n-6)\cdots(n-2k+2)}{(n-2)(n-3)\cdots(n-k+1)}.$$

Proof. Suppose $e = (x, y)$. Suppose i vertices, including x and y, have been chosen into the k-block. Then the number of ways to choose the next vertex is $(n-i)$, and the number of such choices not containing a positive edge is $(n-2i)$ since the other endpoint of the positive edge involving a chosen vertex must be excluded. For example, to choose the next vertex after x and y, not only x and y, but also their positive mates must be excluded. □

Corollary 7.2.12 *For a $(p+1)$-block in a projective plane, $\Phi(n, n/2, p+1) \ge e^{-1/2}(1 + o(1))$.*

Proof.

$$\frac{(n-4)(n-6)\cdots(n-2p)}{(n-2)(n-3)\cdots(n-p)} \ge e^{-(1+o(1))p^2/2n} = e^{-1/2}(1+o(1)).$$

□

7.3 Sequential Algorithms

Consider a set of $n/2$ contigs. The following information-theoretic lower bound was observed in [3, 7].

Theorem 7.3.1 *An algorithm for the contig sequencing problem requires* $0.5n \log n$ *tests asymptotically.*

Proof. There are $(n/2-1)!/2$ distinct Hamiltonian cycles if equivalence is defined by rotations and flipping. Since the test is binary, the number of tests is at least

$$\log[(n/2-1)!/2] \sim 0.5n \log n.$$

□

Grebinski and Kucherov [8] treated contigs as vertices. Their algorithm inserts contigs $1, 2, ..., n/2$ sequentially into a set of chains. At the beginning, the chain is empty. After contig 1 is inserted, there is a single chain consisting of contig 1. After contig 2 is inserted, there is either one chain if contigs 1 and 2 are adjacent, or two if they are not. In general, when contig i is to be inserted, let there be k_i chains, $k_i \leq i-1$. Let L denote the set of left-end contigs of the k chains, and R the set of right-end contigs. Test $L \cup \{i\}$ and $R \cup \{i\}$. If both are negative, then contig i starts a new chain. If one of them is positive, say, the former, and the other not, then use binary splitting to identify the chain whose left-end contigs is adjacent to contig i in $\lceil \log k_i \rceil$ tests, and connect contig i to that chain. If both are positive, then identify the two chains whose left-end and right-end contigs, respectively, are adjacent to contig i, and connect contig i to the two chains to form one chain.

The total number of tests is bounded by

$$\begin{aligned} \sum_{i=1}^{n/2} 2[1+\log k_i] &\leq \sum_{i=1}^{n/2} 2[1+\log(i-1)] \\ &\leq n + n\log(n/2) \\ &= n \log n. \end{aligned}$$

Beigel *et al.* treated primers as vertices in the above algorithm. When primer i, $i = 1, ..., n$, is inserted sequentially into a set S of primers already inserted, we simply test $\{i\} \cup S$. If positive, binary splitting is used to identify the neighbor of i and the positive edge deleted. If negative, just insert i into S. The advantage of using primers instead of contigs is that an identified positive edge can be deleted; so S is always a set of primers with no edge, unlike the contig case.

A test is consumed each time an i is inserted. The test of $\{i\} \cup S$ will be positive in $n/2$ times, and at most $\lceil \log(n/2) \rceil$ tests are needed to identify the positive edge. Hence the total number of tests is bounded by

$$n + (n/2)\log n \to 0.5n \log n \text{ as } n \to \infty.$$

Of course we can also use Johann's algorithm (Section 6.8) to identify the $n/2$ edges in $(n/2)(\lceil \log_2 n \rceil + 7)$ tests. In fact we can even use the Damaschke's algorithm to identify the $n/2$ positive edges one by one in $(n/2)(\lceil \log_2 n \rceil + 1)$ tests since when the defective edges are disjoint, once a defective edge is found, we can delete it by throwing away its two endpoints without worrying of throwing other positive edges away.

The above algorithms are all sequential, i.e., each test takes a round. Hwang and Lin proposed a modification which preserves the asymptotic optimality, but needs only a small expected number of rounds.

Step 1. Find the smallest prime p such that either $p^2 \geq n$ or $p^2 + p + 1 \geq n$. Add $p^2 - n$ (or $p^2 + p + 1 - n$) dummy vertices labeled by $n+1, n+2, ..., p^2$ to obtain an affine (or projective) plane. Remove the dummy vertices and test every block of the plane. Call a block *positive* if its test outcome is positive.

Step 2. Use Damaschke's algorithm to find one edge in each positive block. Remove the edge and test the remaining block. If positive, go to Step 2. If negative, stop.

Note that in Step 1, we first added to the plane dummy vertices and then removed from the block. Thus the sizes of the blocks vary. For example, for vertices $1, 2, ..., 6$, we add 7^*, 8^*, and 9^*, then apply the affine plane method of order $p = 3$ to obtain $p^2 + p = 12$ blocks: $\{1,2,3\}$, $\{4,5,6\}$, $\{7^*,8^*,9^*\}$, $\{1,5,9^*\}$, $\{1,8^*,6\}$, $\{1,4,7^*\}$, $\{4,8^*,3\}$, $\{4,2,9\}$, $\{2,5,8^*\}$, $\{7^*,2,6\}$, $\{7^*,5,3\}$, $\{3,6,9^*\}$. After removing the dummy vertices, we have $\{1,5\}$, $\{1,8\}$, $\{1,4\}$, $\{4,3\}$, $\{4,2\}$, $\{2,5\}$, $\{2,6\}$, $\{5,3\}$, $\{3,6\}$, $\{1,2,3\}$, $\{4,5,6\}$, where nine of them are of size two, and two of them are of size three ($\{7^*,8^*,9^*\}$ no longer exists).

For general n and the smallest prime p such that $p^2 \geq n$, let $z \equiv n \pmod{p}$. Then the design consists of

(1) $(p-z)p$ blocks of size $\lfloor n/p \rfloor$,
(2) zp blocks of size $\lceil n/p \rceil$,
(3) $\lfloor n/p \rfloor$ blocks of size p, and
(4) one block of size z (if z is nonzero).

For convenience, assume that p divides n in the analysis, which means we deal with p^2 blocks of size n/p and n/p blocks of size p in our analysis.

Hwang and Lin showed that for all n there exists a prime p such that $\lceil \log n \rceil \leq \lceil \log p^2 \rceil \leq \lceil \log n \rceil + 1$. Thus in Step 2, the Damaschke's algorithm on B_i takes at most $\lceil \log \binom{p}{2} \rceil \leq \lceil \log n \rceil + 1$ rounds to identify an edge.

To count the number of tests, note that the first round consumes $p(p+1)$ tests and afterwards, each edge requires at most $\log n + 1$ tests to be identified and an additional test to check for any other edge in the remaining block. So it takes a total of $p(p+1) + 0.5n(\log n + 2) \leq 2n + (2n)^{1/2} + 0.5n \log n + n \sim 0.5n \log n$ test for n large. Thus, the algorithm preserves the asymptotic optimality.

By Lemmas 7.2.9 and 7.2.10, we see that for $k = p$ or $p+1$, $P(i, n, k)$ is steeply decreasing in i. Hwang and Lin showed that within three loopings of Step 2, 97.5%

of positive edges are identified. Simulation results showed even better.

7.4 Nonadaptive Algorithms for Matching

Alon *et al.* [1] proved that a 1-round algorithm to identify a matching in K_n requires $O(n^2)$ tests. A few lemmas need to be established towards that proof.

Consider a nonadaptive design matrix M with test sets $\{T\}$ where no test set is contained in another. Treating the n columns as vertices in K_n, M is called *sparse* if there exists a matching on the n vertices such that each T contains at least one edge. Define the weight of M to be

$$w(M) = \sum_{T \in M} 1 \Big/ \binom{|T|+2}{2}.$$

Also define

$$M_{xy} = \{T \setminus \{x,y\} \mid T \in M, \{x,y\} \not\subseteq T\}.$$

Lemma 7.4.1 *Let $S \in M$ be a test set of minimum cardinality $|S| = s \geq 2$. There exists a pair of distinct vertices $\{x,y\} \subseteq S$ such that $w(M_{xy}) \leq w(M)$ if $w(M) \leq 49/153$.*

Proof. It suffices to prove $w(M) \geq E_{\{x,y\} \subseteq S} w(M_{xy})$.

Note that $w(M) \leq 49/153$ implies $|T| \geq 2$ for all $T \in M$, since $|T| = 1$ leads to

$$w(M) \geq 1 \Big/ \binom{1+2}{2} > 49/153.$$

$$\begin{aligned}
& w(M) - E_{\{x,y\} \subseteq S}[w(M_{xy})] \\
= \; & \frac{1}{\binom{s+2}{2}} - \sum_{k<s} \sum_{|T \cap S|=k} \left[\frac{k(s-k)}{\binom{s}{2}} \left(\frac{1}{\binom{|T|+1}{2}} - \frac{1}{\binom{|T|+2}{2}} \right) - \frac{\binom{k}{2}}{\binom{s}{2}\binom{|T|+2}{2}} \right] \\
= \; & \frac{1}{\binom{s+2}{2}} - \frac{1}{\binom{s}{2}} \sum_{k<s} \sum_{|T \cap S|=k} \left[k(s-k) \left(\frac{\binom{|T|+2}{2}}{\binom{|T|+1}{2}} - 1 \right) - \binom{k}{2} \right] \frac{1}{\binom{|T|+2}{2}} \\
= \; & \frac{1}{\binom{s+2}{2}} - \frac{1}{\binom{s}{2}} \sum_{k<s} \sum_{|T \cap S|=k} \left[\frac{2k(s-k)}{|T|} - \binom{k}{2} \right] \frac{1}{\binom{|T|+2}{2}} \\
\geq \; & \frac{1}{\binom{s+2}{2}} - \frac{1}{\binom{s}{2}} \sum_{k<s} \sum_{|T \cap S|=k} \left[\frac{2k(s-k)}{s} - \binom{k}{2} \right] \frac{1}{\binom{|T|+2}{2}} \\
= \; & \frac{1}{\binom{s+2}{2}} - \frac{1}{\binom{s}{2}} \sum_{k<s} f(s,k) \sum_{|T \cap S|=k} \frac{1}{\binom{|T|+2}{2}},
\end{aligned}$$

where $f(s,k) = \frac{2k(s-k)}{s} - \binom{k}{2}$ by definition.

Define $f(s) = \max_{k<s} f(s,k)$. It is easy to verify that

$$f(s) = \max\{f(s,1), f(s,2)\} = \max\{2 - 2/s, 3 - 8/s\}.$$

Therefore

$$\begin{aligned}
& w(M) - E_{\{x,y\}\subseteq S}[w(M_{xy}] \\
\geq\ & \frac{1}{\binom{s+2}{2}} - \frac{1}{\binom{s}{2}} \sum_{k<s} f(s) \sum_{|T\cap S|=k} \frac{1}{\binom{|T|+2}{2}} \\
=\ & \frac{1}{\binom{s+2}{2}} - \frac{f(s)}{\binom{s}{2}} \sum_{k<s} \sum_{|T\cap S|=k} \frac{1}{\binom{|T|+2}{2}} \\
=\ & \frac{1}{\binom{s+2}{2}} - \frac{f(s)}{\binom{s}{2}} \sum_{T\neq S} \frac{1}{\binom{|T|+2}{2}} \\
=\ & \frac{1}{\binom{s+2}{2}} - \frac{f(s)}{\binom{s}{2}} \left[w(M) - \frac{1}{\binom{s+2}{2}} \right].
\end{aligned}$$

Thus, it suffices to prove that

$$\frac{1}{\binom{s+2}{2}} - \frac{f(s)}{\binom{s}{2}} \left[w(M) - \frac{1}{\binom{s+2}{2}} \right] \geq 0,$$

or

$$\frac{1}{\binom{s+2}{2}} \left(\frac{\binom{s}{2}}{f(s)} + 1 \right) \geq w(M).$$

When $f(s) = 2 - 2/s$, the left-hand side $\geq 13/40$ at $s = 3$. When $f(s) = 3 - 8/s$, i.e., $k = 2$ and $s \geq 3$, the left-hand side $\geq 49/153$ at $s = 3$. Since $49/153 < 13/40$, Lemma 7.4.1 is proved. □

Corollary 7.4.2 *M is sparse if $w(M) \leq 49/153$.*

Proof. Repeatedly apply Lemma 7.4.1 to choose disjoint edges (note that the weight remains to be bounded by 49/153) until M is empty, which implies that M is sparse. □

Lemma 7.4.3 *Suppose M identify a matching on n vertices. Then for any two distinct vertices x and y, $\bar{M}_{xy} = \{T \setminus \{x,y\} \mid T \in M$ and $\{x,y\} \subseteq T\}$ is not sparse.*

Suppose $\bar{M}_{xy}$ is sparse for some distinct x and y. Then there is a matching D in the n vertices minus x and y such that each test set in $\bar{M}_{xy}$ contains a positive edge. But the two matchings D and $D \cup \{x,y\}$ yield the same outcome vector, contradicting the assumption that M is (matching) separable. □

Theorem 7.4.4 *Suppose M identifies a matching on n vertices. Then M must have at least $(49/153)\binom{n}{2}$ tests.*

Proof. For each pair $\{x,y\}$ of vertices and each test T containing $\{x,y\}$, count the weight of T after removing $\{x,y\}$. Let W be the sum of those weights over all T and all $\{x,y\}$. There are two ways to count W, grouping according to $\{x,y\}$, or according to T.

(i) Group according to $\{x,y\}$. By Lemma 7.4.3,

$$\sum_{\{x,y\}} w(\bar{M}_{xy}) \geq \sum_{\{x,y\}} 49/153 = (49/153)\binom{n}{2}.$$

(ii) Group according to T.

$$\sum_{T \in M:\{x,y\} \in T} \sum_{\{x,y\}} \frac{1}{\binom{|T|}{2}} = \sum_{T \in M} \binom{|T|}{2} \frac{1}{\binom{|T|}{2}} = |\{T \in M \mid \{x,y\} \in T\}| \leq t.$$

Thus,

$$t \geq (49/153)\binom{n}{2}.$$

□

Note that by testing the $\binom{n}{2}$ edges individually, $\binom{n}{2}$ tests suffice.

In the previous section, we noted that the Steiner design is a decomposition of K_n into cliques and we can work on the subproblem of testing cliques. But decompositions into other subgraphs also work. The following results of Wilson [2] was critically used in [1] to construct a scheme which identifies a matching in K_n.

Theorem 7.4.5 *For any graph G and n large, K_n can be decomposed into edge-disjoint copies of G if and only if $|E(G)|$ divides $\binom{n}{2}$ and gcd (degrees in G) divides $n-1$.*

Since for any graph G, we can always find an infinite class of n satisfying those conditions, the problem is reduced to finding a G with a good $t(G)/|E(G)|$ ratio, and then use $t(G)$ as a subroutine for $t(K_n)$. Namely,

Theorem 7.4.6

$$t(K_n) \leq \frac{t(G)}{|E(G)|}\binom{n}{2} + O(n).$$

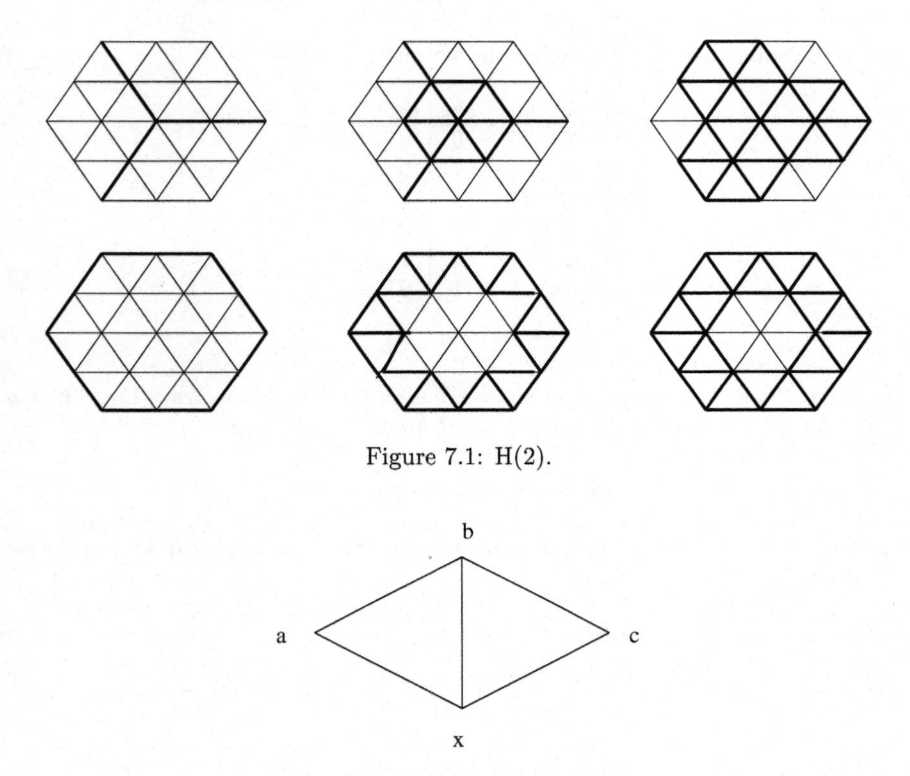

Figure 7.1: H(2).

Figure 7.2: Two adjacent triangles.

Note that a test in G cannot contain an edge in $K_n \setminus G$ or it becomes a different test when G is a subgraph of K_n. Therefore, any test in G must be on a clique.

Before we give the currently best result, we first give an intermediate result.

Let $H(s)$ be a triangulated hexagon of side s consisting of regular triangles of unit edge-length (see Fig. 7.1).

It is easily verified that $H(s)$ has $3s^2 + 3s + 1$ vertices, $9s^2 + 3s$ edges and $6s^2$ triangles.

Lemma 7.4.7 *If we know the status of the subsets* $\{a, b\}$, $\{b, c\}$, $\{a, b, x\}$ *and* $\{b, c, x\}$ *in Fig. 7.2, then we know the status of* $\{a, x\}$, $\{b, x\}$, $\{c, x\}$.

Proof. We enumerate all cases except the symmetrical ones.

ab	bc	abx	bcx	ax	bx	cx
0	0	0	0	0	0	0
0	0	0	1	0	0	1
0	0	1	1	0	1	0
0	1	0	1	0	0	0
0	1	1	1	1	0	0
1	1	1	1	0	0	0

□

Corollary 7.4.8 *It suffices to test all* $3s$ *edges on the three axes (heavy lines) in Fig. 7.1 and the* $6s^2$ *triangles to identify the matching.*

Note that the identification is from center to boundary.

Corollary 7.4.9 *It suffices to test all* $6s$ *edges on the boundary of* $H(s)$ *and the* $6s^2$ *triangles to identify the matching.*

Note that the identification is from boundary to center.
Either corollary will yield the result

$$\frac{t(H(s))}{|E(H(s))|} = \frac{2}{3} + o(s^{-1}).$$

Although the algorithm in Corollary 7.4.9 is slightly worse than the one in Corollary 7.4.8, it will be the one used later in some extension due to the following nice property.

Corollary 7.4.10 *Suppose the status of a set* T *of triangles are not known. The status of an edge can still be determined if it lies in a triangle which can be accessed from the boundary without going through any triangle in* T.

Let H^+ be the graph obtained from $H(s)$ by adding a vertex v and $3s^2 + 3s + 1$ edges (v, u), $u \in H(s)$. Let M consist of the following two types of tests:
(i) tetrahedra: test $v \cup \Delta$ for every $\Delta \in H(s)$.
(ii) boundary: test every boundary edge of $H(s)$ and vu for every boundary vertex u.

We now show that these tests suffice to identify the matching. If edge (vu) is in the matching for some interior vertex u, then the six tetrahedra containing (vu) must be all positive. We call an interior vertex *super* if all its six tetrahedra are positive. It is easily seen

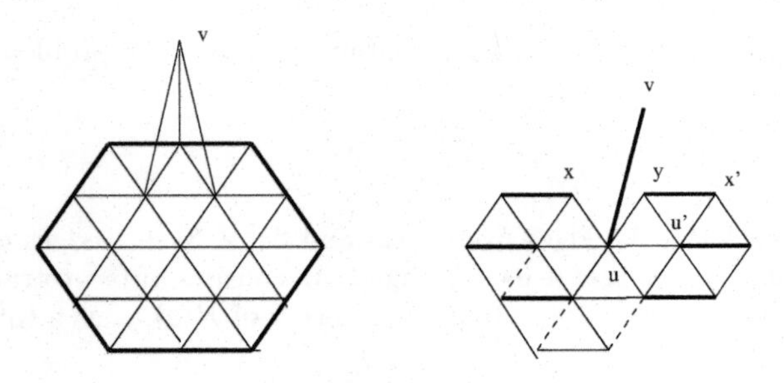

Figure 7.3: Two super vertices.

Lemma 7.4.11 *u is super if and only if either*

(i) (vu) is in the matching, or

(ii) (vu') is in the matching, u is adjacent to u' and the three heavy edges in Fig. 7.3 are in the matching.

Let k denote the number of super vertices in $H(s)$.

Corollary 7.4.12 $k \leq 3$.

Theorem 7.4.13 *M identifies the matching.*

Proof. By Corollary 7.4.12, it suffices to consider the following cases:

(i) $k = 0$. If (vu) is in the matching, u must be a boundary vertex, which will be identified by a boundary test. All tetrahedra tests can be converted to triangle tests on $H(s)$ except the tetrahedra containing (vu) but no boundary edge. However, all three edges of the corresponding triangle also belong to other triangles and thus their status can be identified by Corollary 7.4.12.

(ii) $k = 1$. Then (vu) is in the matching and u is the unique super vertex. All tetrahedra tests are converted to triangle tests except the six tetrahedra containing (vu) whose corresponding triangles forming a $H(1)$. Since its six interior edges all involve u, only the six boundary edges of $H(1)$ can be in the matching. But these six boundary edges all lie in some other triangles and their status can be identified by Corollary 7.4.10.

(iii) $k = 2$. By Lemma 7.4.12, the two super vertices u and u' must be adjacent. Suppose (vu') is in the matching. Then (xy) must also be in the matching (see Fig. 7.3), and hence $(x'y)$ is not. All tetrahedra tests can be converted to triangle tests except the ten tetrahedra containing either (vu) or (vu'). By Corollary 7.4.10, the status of the eight boundary edges in Fig. 7.3 can be identified. Without loss of generality, suppose (xy) is in the matching and $(x'y)$ not, then (vu') and (zu) are in the matching but not the other ten interior edges in Fig. 7.3.

(iv) $k = 3$. Then the three super vertices are adjacent on a line with the middle one being the u with (vu) in the matching. All tetrahedra tests can be converted to triangle tests except the six tetrahedra containing (vu). The problem is reduced to the case studied in $k = 1$. □

Corollary 7.4.14 $t(K_n) \leq \frac{1}{2}\binom{n}{2} + O(n)$.

Proof. $H^+(s)$ has $9s^2 + 3s$ edges from $H(s)$ and $3s^2 + 3s + 1$ edges of the type vu, $u \in H(s)$, thus a total of $12s^2 + 6s + 1$ edges. M consists of $6s^2$ tetrahedra tests, $6s$ boundary tests of the type vu, is a boundary vertex of $H(s)$, thus a total of $6s^2 + 12s$ tests.

$$\frac{t(H^+(s))}{|E(H^+(s))|} = \frac{6s^2 + 12s}{12s^2 + 6s + 1} = \frac{1}{2} + o(s^{-1}).$$

Corollary 7.4.14 now follows from Theorem 7.4.6. □

When we relax 1 round to k rounds, $O(n^2)$ can be improved. Alon *et al.* gave a 2-round algorithm.

Theorem 7.4.15 *There is a 2-round algorithm to identify a matching in K_n in about $(5/4)n^{3/2}$ tests asymptotically.*

Proof. Decompose K_n into approximately $n^{3/2}$ copies of K_m where $m = n^{1/4}$. Test each copy of K_m. At most $n/2$ of them are positive. Test each edge in a positive copy. The total number of tests is at most

$$\binom{n}{2} \Big/ \binom{n/4}{2} + \frac{n}{2}\binom{n/4}{2} = \frac{5}{4}n^{3/2} + O(n).$$

□

We can extend Theorem 7.4.15 to k-round algorithms.

Theorem 7.4.16 *There exists a k-round algorithm for $k \geq 2$ to identify a matching in K_n with $O(n^{1+1/k}\log n)$ tests.*

Proof. The $k = 2$ case was proved in Theorem 7.4.15. For $k \geq 3$, we prove Theorem 7.4.16 by induction on k.

Decompose K_n into approximately $n^{2-2(k-1)/k^2}$ copies of K_m where $m = (k-1)/k^2$. Test each copy; at most $n/2$ of them are positive. Use induction to identify all positive edges in a positive copy. The total number of tests is at most

$$n^{2-2(k-1)/k^2} + \frac{n}{2}(n^{(k-1)/k^2})^{1+1/(k-1)}\log n = O(n^{1+1/k}\log n).$$

□

Alon *et al.* considered the bipartite graph $G(L, R, E)$ where $|R| \geq 2$.

Lemma 7.4.17 *There exists a k-round algorithm, $k \geq 1$, to identify a matching M in $G(L, R, E)$ with $|M||L|^{1/k} \log |R|$ tests per round.*

Proof. For $k = 1$ we can identify the matching-edge, if it exists, from a vertex of L to R in $\log |R|$ tests. Hence $|L| \log |R|$ tests suffice to identify the matching. We prove the $k = 2$ case by induction on k.

Partition L into $t = |L|^{1/k}$ pieces $L_1, ..., L_t$ of size $|L|/t$. Test $L_i \cup R$ for $1 \leq i \leq t$. Then the first round takes $|L|^{1/k}$ tests. At most $|M|$ such tests are positive. Let m_i denote the number of matching edges in the ith piece. By the induction hypothesis, there exists a $(k-1)$-round algorithm which identifies the matching edges in all positive $L_i \cup R$ in

$$\sum_i m_i(|L|/t)^{1/(k-1)} \log |R| = |M||L|^{1/k} \log |R|$$

tests per round. □

Corollary 7.4.18 *For $1 > c > |L|^{-1/k}$, there exists a k-round algorithm, $k \geq 2$, to identify a matching M in $G(L, R, E)$ where the first round takes at most $c|L|^{1/k}$ tests and each subsequent round takes at most $|M||L|^{1/k} \log |R|/c^{1/k}$ tests.*

Proof. Partition L into $t = c|L|^{1/k}$ pieces $L_1, ..., L_t$ of size $|L|/t$ and do the same as in Lemma 7.4.7. In particular, apply the algorithm there to each positive $L_i \cup R$. □

Alon *et al.* used the bipartite graph results to improve on Theorem 7.4.16. This implies that K_n is to be converted into bipartite graphs preserving all edges.

First, consider a lemma from Alon and Spencer [2]. Let $N(v)$ denote the set of neighbors of a vertex v.

Lemma 7.4.19 *Let $B = (U, V, E)$ be a d-regular bipartite graph where $|U| = |V| = n$. Let A be the adjacent matrix of B. Suppose every eigenvalue of $A^T A$ except the largest one (which is d^2) is at most λ^2. Then for every $S \subseteq V$.*

$$\sum_{u \in U} (|N(u) \cap S| - d\frac{|S|}{n})^2 \leq \lambda^2 |S|(1 - \frac{|S|}{n}),$$

where $N(u)$ is the set of neighbors of u.

Lemma 7.4.20 *Let P be a finite projective plane with $n = p^2 + p + 1$ points. Let Q be obtained by deleting at least $n/2$ lines of P. Let $G = (V, E)$ where V is P's point-set and E consists of all pairs (x, y) such that there is a line in Q containing both x and y. Then G is $\sqrt{n} \log n(1 + o(1))$ colorable using color classes of sizes at most $\sqrt{n}$. Further, such a coloring can be found in polynomial time.*

Proof. Define $B(U,V,E')$ such that V is the vertex-set of G, U is the set of lines of P, and $uv \in E'$ if v is incident to u in P. Let A be adjacency matrix of B with U as columns. Then A^TA has all entries 1 except the diagonal entries are $p+1$. Consequently, the largest eigenvalue of A^TA is $(p+1)^2$ and all other eigenvalues are p. Let $S \subset V$. By Lemma 7.4.19

$$\sum_{u\in U}(|N(u)\cap S| - d\frac{|S|}{n})^2 \leq p|S|(1-\frac{|S|}{n}).$$

Let ℓ be the number of u with $|N(u)\cap S| < \frac{d|S|}{2n}$. Then

$$\ell(\frac{d|S|}{2n})^2 \leq \sum_{u\in U}(|N(u)\cap S| - d\frac{|S|}{n})^2 \leq p|S|(1-\frac{|S|}{n}) \leq p|S|,$$

or

$$\ell \leq \frac{4pn^2}{d^2|S|} \leq \frac{4n^{3/2}}{|S|}.$$

By setting $|S| > 10\sqrt{n}$, every set consisting of $0.4n$ lines contains at least one line which contains at least $(d+1)|S|/(2n) \geq \sqrt{n}|S|/(2n) = |S|/(2\sqrt{n})$ points of S.

This argument will be used iteratively on a set L consisting of at least $0.4n$ lines. Initially, let $S = V$. Then we may choose a line in L which contains at least a $1/(2\sqrt{n})$ fraction of the points in S. Remove those points from S. After at most $2\sqrt{n}\ln(n/10\sqrt{n}) < \sqrt{n}\ln n$ iterations, S is reduced to size at most $10\sqrt{n}$.

To complete the proof, set L to be the set of lines deleted from P. Then the color classes are the $\sqrt{n}\ln n$ lines chosen in the iterations plus the $10\sqrt{n}$ points (each as a color class) remained in S. If a point belongs to more than one color class, choose one arbitrarily. □

We are now ready to prove the main result.

Theorem 7.4.21 *For $2 \leq k \leq \log n$, there is a deterministic k-round algorithm to identify a matching in K_n with $O(n^{1+1/2(k-1)}(\log n)^{1+1/(k-1)})$ tests per round.*

Proof. In the first round, test each line in P. At most $d \leq n/2$ of them contain a positive edge. Set L to be those lines containing no positive edge which derive the $O(\sqrt{n}\log n)$ color classes. Note that no color class can contain an edge. For every pair (X,Y) of color classes, $|Y| \geq 2$, apply the $(k-1)$-round algorithm from Corollary 7.4.18.

Set $c = (\log n)^{1/(k-1)-1}$. Then the second round takes

$$\binom{\sqrt{n}\log n}{2} c\sqrt{n}^{1/(k-1)} = O(n^{1+1/2(k-1)}(\log n)^{1+1/(k-1)})$$

tests. Apply Corollary 7.4.18, each subsequent round takes

$$\frac{n}{2}\sqrt{n}^{1/(k-1)} \log \sqrt{n}/c^{1/(k-1)} = O(n^{1+1/2(k-1)}(\log n)^{1+1/(k-1)})$$

tests. □

Alon *et al.* found that a random algorithm can be much more effective. Randomly permute the n points to obtain a random projective plane. Test each line. By Corollary 7.2.12, each negative edge has probability at least $e^{-1/2}(1+o(1))$ of being identified.

Do the permutation $m \log n$ times to obtain $m \log n$ random projective plane. Then the probability that a negative edge is not identified is at most

$$\pi(m) = (1 - e^{-1/2})^{m \log n}(1 + o(1)).$$

If m is set to be $(1+o(1)) \ln 2/\ln(1/(1-e^{-1/2})) \sim 0.74$, then $\pi(m) \leq 1/n$. Thus, the expected number of negative edges not identified in these tests is at most

$$\binom{n}{2}/n = n/2.$$

Hence, the number of unidentified negative edges and the $d \leq n/2$ positive edges can be identified in the second round in at most n tests, leading to a total of $0.74n \log n + n$ tests.

The second round can be eliminated by choosing m twice as large plus ϵ. Then $\pi(d) = o(1/n^2)$, and the 1-round algorithm identifies all matches with high probability in approximately $(2 \times 0.74 + \epsilon)n \log n \sim 1.49n \log n$ tests. They also stated that the 2-round algorithm can be improved to $0.72n \log n$ tests, and the 1-round to $1.44n \log n$.

Beigel *et al.* proposed a more sophisticated algorithm which has three steps. The first step partitions the n vertices into disjoint groups and positive edges within a group are identified. The second step identifies the inter-group positive edges by first checking which pairs of group test positive and then use this information to generate a candidate set of positive edges for confirmation tests in the third step. To reduce the number of confirmation tests, the partitioning part is actually done many times to obtain more groups, hence more inter-group information. Note that one can use a bipartite graph algorithm to identify all positive edges between two groups, but that would consume too many tests.

We now give their algorithm in detail. Partition the n vertices into $n^{0.7}$ pieces of $n^{0.3}$ vertices each. Test each of the $n^{0.7}$ pieces; if positive, identify and remove all positive edges in the piece by testing all $n^{0.3}$ pairs of vertices. By Corollary 7.2.2, the expected number of tests consumed is

$$n^{0.7} + 0.5n^{0.3}\binom{n^{0.3}}{2} \leq n^{0.7} + 0.25n^{0.9}.$$

In the next step partition the $n^{0.7}$ pieces into $n^{0.5}/\sqrt{c}$ groups of $\sqrt{c}n^{0.2}$ pieces each, where c is a constant to be optimized. Test each group. If positive, identify all its positive pairs of pieces. For each such pair, identify and remove its positive edges by using the 1-round bipartite graph algorithm given in the proof of Theorem 7.4.21. By Corollaries 7.2.2 and 7.2.3, the expected number of positive groups and the expected number of positive edges are both at most $0.5\sqrt{c}n^{0.5}$. Hence the number of tests consumed at this step is

$$\begin{aligned} & n^{0.5}/\sqrt{c} + 0.5\sqrt{c}n^{0.5}\binom{\sqrt{c}n^{0.2}}{2} + 0.5\sqrt{c}n^{0.5}(n^{0.3}\log n^{0.3}) \\ = \; & n^{0.5}/\sqrt{c} + 0.25\sqrt{c}^3 n^{0.9} + 0.15\sqrt{c}n^{0.8}\log n. \end{aligned}$$

Adding the tests in the two steps, and do the partition procedure $\sqrt{\log n}$ times, the total number of tests is

$$[n^{0.7} + 0.25(1+\sqrt{c}^3)n^{0.9} + n^{0.8}\log n]\sqrt{\log n}.$$

The $\sqrt{\log n}$ partitions yield $\sqrt{n\log n/c}$ groups with no internal edges. Test every pair of group which take

$$\binom{\sqrt{n\log n/c}}{2} \sim 0.5n\log n/c \text{ tests.}$$

Consider a pair (i,j) of vertices. If there exists a group containing both i and j, then whether (i,j) is a positive edge is established in the first two steps. So assume i and j never appear together in a group. If there exist two groups A and B such that $i \in A$, $j \in B$ and the test on $A \cup B$ is negative, then (i,j) cannot be a positive edge. If the test on every $A \cup B$ with $i \in A$ and $j \in B$ is positive, then (i,j) is tested for confirmation to be a positive edge. While every positive edge not within a group winds up with the confirmation test, Beigel *et al.* showed that the probability of a negative edge to take a confirmation test is at most $[f(\log n, c)]/n$ for some function f. Hence the expected number of confirmation tests is at most

$$n/2 + \binom{n}{2} f(\log n, c)/n \sim n/2 f(\log n, c).$$

Similarly, the total number of tests is

$$\begin{aligned} & (n^{0.7} + 0.25(1+\sqrt{c}^3)n^{0.9} + n^{0.8}\log n)\sqrt{\log n} + 0.5n\log n/c + n/2f(\log n, c) \\ \to \; & \frac{0.5}{c} n\log n + \frac{n}{2} f(\log n, c). \end{aligned}$$

The minimum $0.72135n\log n$ is obtained by setting $c = \ln 2$ (see Fig. 7.4).

It takes one round to test the pieces and another to identify positive edges in the positive pieces. A third round is taken to test the groups, a fourth to test all pairs of pieces in the positive groups, and a fifth to identify all positive edges in the positive pairs of pieces. Finally, the test of all pairs of groups takes round six and the confirmation tests takes round seven.

---

item: 1, ..., $n^{0.3}$, ..., n

piece: $\boxed{P_1}$ $\cdots$ $\boxed{P_{\sqrt{c}n^{0.2}}}$ $\cdots$ $\boxed{P_{n^{0.7}}}$

Test every piece; if positive, test all pairs

$$n^{0.7} + \frac{n^{0.3}}{2}\binom{n^{0.3}}{2} \sim \frac{1}{4}n^{0.9}$$

group: $\boxed{G_1}$ $\cdots$ $\boxed{G_{\sqrt{n/c}}}$

Test every group; if positive, identify all positive pairs of pices; for each positive pair, identify all positive edges

$$\sqrt{n/c} + \frac{\sqrt{nc}}{2}\binom{\sqrt{c}n^{0.2}}{2} + \frac{\sqrt{nc}}{2}n^{0.3}\log n^{0.3} \sim \frac{c^{3/2}}{4}n^{0.9},$$

repeat $\sqrt{\log n}$ times; yielding $\sqrt{n \log n/c}$ groups.

Test every pair of groups

$$\binom{\sqrt{n \log n/c}}{2} \sim n \log n/c.$$

Note that all edges in a group are identified. Further, if a pair of groups tests negative, then all inter-group edges are negative. All edges not identified are tested individually. This number is shown to be $\alpha cn \log n$ where α is a constant. The minimum $0.72135n \log n$ is obtained by setting $c = \ln 2$.

Figure 7.4: $0.72135n \log n$ is achieved by setting $c = \ln 2$.

7.5 The 3-Stage Procedure

Grebinski and Kucherov [7] considered the quantitative model in which a test reveals the number of positive edges in the subset tested. We first cite some relevant literature.

A binary matrix M is called a *detecting matrix* if for any two distinct integral vectors u_1 and u_2, $Mu_1 \neq Mu_2$, namely, arithmetic sum of different sets of columns of M are different. M is *q-detecting* if the vectors u_i are restricted to q-nary vectors, and *(q, d)-detecting* if u_i is further restricted to at most d nonzero element.

Theorem 7.5.1 *A $k \times n$ 2-detecting matrix can be constructed with $k = 2n/\log n$ tests asymptotically.*

Erdös and Renyi [6] proved that $2n/\log n$ is an asymptotic upper bound and conjectured its realizability, which was proved by Lindstrom [10] and Cantor and Mills [4].

Grebinski and Kucherov [7] extended Theorem 7.5.1 to 3-detecting.

Theorem 7.5.2 *A $k \times n$ ternary detecting matrix can be constructed with $k = 4n/\log n$ tests asymptotically.*

Lindstrom [10] also proved

Theorem 7.5.3 *A $k \times n$ (2,2)-detecting matrix can be constructed with $k = 2\log n$ tests asymptotically.*

When $G = K_{mn}$, a complete bipartite graph with m and n vertices in the two respective parts, Grebinski and Kucherov proved

Theorem 7.5.4 *There exists a nonadaptive algorithm to identify a subgraph with maximum degree 2 in G by $8m\log n/\log m$ tests if each of the m vertices has maximum degree 2.*

Proof. As shown in Fig. 7.5, take a $4m/\log m \times m$ 3-detecting matrix M and a $2\log n \times n$ (2,2)-detecting matrix N to obtain a $(4m/\log m)(2\log n) \times (m+n)$ matrix $M * N$ where row (i, i') is the concatenation of row i of M and row i' of N, $1 \leq i \leq 4m/\log m$, $1 \leq i' \leq 2\log n$. By collecting the outcomes of tests involving row i' of N, we learn the number of neighbors (0, 1 or 2) of vertex (column in M) $j \notin U$ in row i' since M is a 3-detecting matrix. On the other hand, for a fixed j, since we know the number of its neighbors in row i' of N for any i', and N is a 2-separating matrix, we can identify its two neighbors in V (columns in N). Note that the total number of tests is about

$$(4m/\log m)(2\log n) = 8m\log n/\log m.$$

□

$\frac{4m}{\log m} \times m$ M: 3-detecting

$$\begin{array}{cc} & v_1 \; v_2 \; \cdots \; v_m \\ 1 & \text{row } 1 \\ \vdots & \vdots \\ i & \text{row } i \\ \vdots & \vdots \\ s & \text{row } s \end{array}$$

where $s = \frac{4m}{\log m}$.

$2 \log n \times n$ N: (2,2)-detecting

$$\begin{array}{cc} & v'_1 \; v'_2 \; \cdots \; v'_m \\ 1' & \text{row } 1' \\ \vdots & \vdots \\ i' & \text{row } i' \\ \vdots & \vdots \\ t' & \text{row } t' \end{array}$$

where $t = 2 \log n$.

$\frac{8m \log n}{\log m} \times (m+n)$ $M * N$:

$$\begin{array}{ccc} & v_1 \; v_2 \; \cdots \; v_m & v'_1 \; v'_2 \; \cdots \; v'_n \\ (1, 1') & \text{row } 1 & \text{row } 1' \\ \vdots & \vdots & \vdots \\ (1, t') & \text{row } 1 & \text{row } t' \\ \vdots & \vdots & \vdots \\ (i, i') & \text{row } i & \text{row } i' \\ \vdots & \vdots & \vdots \\ (s, t') & \text{row } s & \text{row } t' \end{array}$$

Figure 7.5: Construction of $M * N$.

Corollary 7.5.5 *$8n$ tests suffice if $m = n$.*

Grebinski and Kucherov [7] observed the following information-theoretic lower bound.

Lemma 7.5.6 *The number of tests to identify a Hamiltonian cycle in K_n under the quantitative multi-vertex model is at least $\log_{n/2+1}(n-1)!/2 \to n$ asymptotically.*

Proof. $(n-1)!/2$ is the number of nonequivalent Hamiltonian cycles, while each test has $n/2+1$ possible outcomes (the number of positive edges it contains). Note that

$$\begin{aligned} \log_{n/2+1}(n-1)!/2 &\sim (n-1)\log_{n/2+1}(n-1) \\ &\sim n\log_{n/2+1} 2(n/2) \\ &\sim n \text{ as } n \to \infty. \end{aligned}$$

□

They also gave an $O(n)$-test sequential algorithm. The trick is to break the complete graph G into complete bipartite graphs such that Lemma 7.5.4 can apply. At first, they proposed a sequential algorithm for this breaking up. The $n/2$ vertices are partitioned into three subsets A, B, C such that no positive edge exists within a subset. This can be done by processing the vertex one by one. Suppose $i-1$ vertices have been assigned and we are processing vertex i. Test $i \cup A$. If negative, assign i to A; if positive, test $i \cup B$. If negative, assign i to B. Otherwise, assign i to C (since we know i has at most two neighbors, three parts suffice). It takes at most n tests to do the partition.

It is easily verified that $|A| \leq n/4$, $|B| \leq n/4$ and $|C| \leq n/4$. For each of the three bipartite graphs on (A, B), (A, C), (B, C), respectively, apply Corollary 7.5.5 on $K_{n/4,n/4}$ to obtain $8(n/4) = 2n$ tests. Thus, the total number of tests is

$$n + 3(2n) = 7n.$$

Hwang and Lin [9] proposed to use primers as vertices. Since a primer has at most one neighbor, the requirement on M can be reduced from 3-detecting to 2-detecting in Theorem 7.5.4. This reduces $8n$ tests to $2n$ tests. Further, the vertices need to break into only two parts with a single bipartite graph $K_{n/2,n/2}$. The partition takes n tests.

Note that there are n primers instead of $n/2$ contigs. Hence, the bipartite graph in concern is $K_{n/2,n/2}$ and the number of tests in $M * N$ is $2(n/2) = n$ tests. Hence the total number of tests is

$$n + n = 2n.$$

Grebinski and Kucherov [8] also found a way to get rid of the sequential part by not doing the partition of vertices. Construct a complete bipartite graph H with vertex-sets A and B where $A = B = V$, and each positive edge (u,v), $u < v$, in G yields a positive edge (u,v), $u \in A$, $v \in B$, in H. Apply the nonadaptive algorithm on H.

Note that the bipartite graph H is an artificial creation. A test on $X \cup Y$, $X \subseteq A$ and $Y \subseteq B$, is not just on the bipartite subgraph $H|_{(X,Y)}$ induced by X and Y, but also on the edges within X and within Y. Since the nonadaptive algorithm on a bipartite graph tests only the bipartite part, we need to filter out the "within" part. Grebinski and Kucherov gave a formula to convert each artificial test $\mu'(X,Y)$ to a set of real tests μ (on distinct edges):

$$\mu'(X,Y) = \mu[(X \setminus Y) \cup (Y \setminus X)] - 2\mu(X \setminus Y) - 2\mu(Y \setminus X) + \mu(X) + \mu(Y).$$

Thus, each artificial test is replaced by five real tests. A further reduction on the number of tests is possible. Note that there are only $2n/\log n$ varieties of X and $\log n$ varieties of Y, hence $\mu(X)$, $\mu(Y)$ can be computed first from all X, Y in $O(n/\log n + \log n)$ time. Then each artificial test needs the computation of only three real tests, leading to

$$O(n/\log n + \log n) + 7n \times 3 = 21n + o(n)$$

as the total (asymptotic) number of tests.

Again, Hwang and Lin proposed primers as vertices. Then the number of tests is $2n \times 3 + o(n) = 6n + o(n)$.

Hwang and Lin proposed the following algorithm for the incremental model with primers as vertices. Let M denote the binary representative matrix with $n-1$ columns which was shown to be 1-separable in Chapter 2.

1. Let M^j ($M^{(j)}$) be obtained from M by adding a column of 1's (0's) as column j.

2. By comparing the outcomes of test i, $i = 1, ..., t$, in M^j and $M^{(j)}$, we find out whether test i in M contains a neighbor of node j.

3. Since M is 1-separable, the outcomes of the t tests in M suffice to identify the neighbor of node j.

We can do the tests in M^j and $M^{(j)}$ for each $j = 1, ..., n$. But, we can do better by recognizing duplicated tests in the sets $\{M^j\}$ and $\{M^{(j)}\}$.

Consider the n spaces separated by the $n-1$ columns of M. Let M' be considered by adding a column of 1's at space j. A space is called a *1-space* (or *0-space*) if it neighbors a 1 (0). Note that a space can be both a 1-space and a 0-space. If a test contains a run of k 1's, then adding 1 at any one of the $k+1$ spaces separated by the k consecutive 1's induces the same test. Suppose test i of M has x 1's and r runs of 1's. Then it has $x+r$ 1-spaces, which induce r distinct tests. Thus, the total number of distinct tests in $\{M^j\}$ corresponding to test i is $n - (x+r) + r = n - x$. On the other hand, using a similar argument, the total number of distinct tests in $\{M^{(j)}\}$ corresponding to test i is $n - (n-1-x) = x+1$ where $n-1-x$ is the number of 0's in test i. Summing up, a test in M induces $n+1$ distinct tests. So $\{M^j\}$ and $\{M^{(j)}\}$ together induces $(n+1)t = (n+1)\log n$ distinct tests.

Note that this is a nonadaptive algorithm with number of tests doubling the asymptotic bound of sequential algorithms under the multi-vertex model.

References

[1] N. Alon, R. Beigel, S. Kasif, S. Rudich and B. Sudakov, Learning a hidden matching, *SIAM J. Comput.*, 33 (2004) 487-501.

[2] N. Alon and J. H. Spencer, *The probabilistic Method*, Second Edition, Wiley, New York, 2000.

[3] R. Beigel, N. Alon, M. S. Aprydin, L. Fortnow and S. Kasif, An optimal procedure for gap closing in whole genome shotgun sequencing, *Proc. RECOMB*, ACM Press, 2001, pp. 22-30.

[4] D. G. Cantor and W. H. Mills, Determination of a subset from some combinatorial properties, *Can. J. Math.*, 18 (1966) 42-48.

[5] D. Damaschke, A tight upper bound for group testing in graphs, *Disc. Appl. Math.*, 48 (1994) 101-109.

[6] P. Erdös and A. Rényi, On two problems of information theory, *Bubl. Hungarian Acad. Sci.*, 8 (1963) 241-254.

[7] V. Grebinski and G. Kucherov, Reconstructing a Hamiltonian cycle by querying the graph: Application to DNA physical mapping, *Disc. Appl. Math.*, 88 (1998) 147-165.

[8] V. Grebinski and G. Kucherov, Optimal reconstruction of graphs under the additive model, *Algorithmica*, 25 (2000) 104-124.

[9] F. K. Hwang and W. D. Lin, The incremental group testing model for gap closing in sequencing long molecules, *J. Combin. Opt.* 7 (2003) 327-337.

[10] B. Lindstiöm, On a combinatorial problem in number theory, *Canad. J. Math.*, 8 (1965) 402-407.

[11] B. Lindstrom, Determination of two vertices from a sum, *J. Combin. Thy.*, A6 (1969) 402-407.

[12] H. Tettelin, D. Radune, S. Kasif, H. Khousi and S. L. Salzberg, Optimized multiplex PCR: Efficiently closing a whole genome shotgun sequencing project, *Genomics* 62 (1999) 500-507.

[13] R. M. Wilson, Decomposition of complete graphs into subgraphs isomorphic to a given graph, *Congressus Numeratium*, XV (1975) 647-659.

8
The Inhibitor Model

In some biological applications, a clone can be positive, negative, or anti-positive in the sense that anti-positives cancel with the positives. They are called *inhibitors* in the literature. Different models can be formulated by weighing the cancellation effect differently.

8.1 Introduction

In the simplest model, first proposed by Farach *et al.* [9] and called the *inhibitor model*, the presence of a single inhibitor in a pool distates the outcome to be negative. In particular, if there are up to d positive clones and up to s inhibitors, then we called it a $(\bar{d}, \bar{s})$ inhibitor model. The usual concern in the inhibitor model is to identify all positives in the presence of inhibitors. A secondary problem one can also consider is to also identify the inhibitors (hence the negatives too). Farach *et al.* showed that the two problems are closely related.

Theorem 8.1.1 *Suppose the positives are identified in t tests. Then the inhibitors can be identified in* $\min\{t, s\log n\}$ *sequential tests, or* $\min\{t, O(s^2 \log n)\}$ *nonadaptive tests, where s is the number of inhibitors.*

Proof. Consider the set of nonpositive clones. Let S be an arbitrary subset and D_1 an identified positive clone. Then S contains no inhibitor if and only if the test outcome on $D_1 \cup S$ is positive. Mathematically, this is equivalent to the group testing mechanism where a set contains no positive if and only if the test outcome is negative. Therefore by interpreting inhibitors as "positives" and positive outcome as "negative outcomes", a group testing algorithm (with $s \log n$ sequential tests and $O(s^2 \log n)$ noadaptive tests) can identify all inhibitors.

Next we show t more tests suffice. Let U denote the set of clones whose states are uncertain at the time when all positives are identified. Namely, U consists of clones never appearing in a positive test. Then there is no $C \in U$ such that every pool containing C also contains another clone in U; otherwise, C could be a positive with those other columns as inhibitors, contradicting the assumption that all positives are identified. In other words, thre exists a pool T_C (with a negative outcome) containing C as the only member of U. This means that $|U| \leq t$. Let D_1 denote an identified positive clone. Then testing $D_1 \cup \{C\}$ for every $C \in U$ reveals whether C is an inhibitor. □

Theorem 8.1.2 *The information-theoretic lower bound for identifying all d positives and all s inhibitors is $O((d+s)\log n)$ for $s, d << n$.*

Proof. The number of possible distributions of the d positives and the r inhibitors among the n clones is

$$\frac{n!}{d!s!(n-d-s)!} \sim n^{d+s}.$$

□

De Bonis and Vaccaro [3] gave a lower bound for the nonadaptive case.

Theorem 8.1.3 *A pooling design for the (d, s) inhibitor model is a (d, s)-disjunct matrix.*

Proof. Let M be a pooling design for the (d, s) inhibitor model. Suppose M is not (d, s)-disjunct. Then there exists a set of $d+s$ columns $C_1, ..., C_{d+s}$ such that $\cup_{i=1}^{d} C_i \subseteq \cup_{i=d+1}^{d+s} C_i$. Consider the sample point that $\{C_1, ..., C_d\}$ is the set of positive items and $\{C_{d+1}, ..., C_{d+s}\}$ is the set of inhibitors. Then all outcomes are negative, and we cannot distinguish $\{C_1, ..., C_d\}$ from the empty set for D. □

Thus, $t(d, s, n)$ studied in Section 2.6 is a lower bound of a pooling design for the (d, s) inhibitor model.

The history of the inhibitor model goes through the following steps:

(i) Find an algorithm to show that the problem is solvable. At first, only a nondeterministic sequential algorithm was given.

(ii) The nondeterministic sequential algorithm was improved to a deterministic one.

(iii) Some k-rounds algorithms were proposed for some k.

(iv) A 1-round algorithm was proposed.

(v) An error-tolerant 1-round algorithm was proposed.

(vi) Extension to other inhibitor models.

However, our presentation will not be dictated by this historical order. In fact, we will jump right to the final solution, i.e., step (v).

8.2 1-Round Algorithm

Dýachkov, Macula, Torney and Villenkin [6] first gave a 1-round design by proving

Theorem 8.2.1 *A $(d+s)$-disjunct matrix identifies all positive clones in a $(\bar{d}, \bar{s})$ inhibitor model.*

Hwang and Liu [10] extended this result to the error-tolerant case.

Theorem 8.2.2 *A $(d+s+2e)$-disjunct matrix identifies all positive clones in a $(\bar{d}, \bar{s})$ inhibitor model with at most e errors.*

We will present an improved version.

Theorem 8.2.3 *A $(d+s; 2e+1)$-disjunct matrix identifies all positive clones in a $(\bar{d}, \bar{s})$ inhibitor model with at most e errors.*

Before we prove Theorem 8.2.3, we derive some basic facts. For a nonisolated test matrix M and a given outcome vector V, let $t_0(C)$ $(t_1(C))$ denote the number of negative (positive) tests containing C. Let P, Q, I denote a positive, a negative and an inhibitor clone, respectively.

Lemma 8.2.4 *Suppose M is $(m; z)$-disjunct with $m \geq s$. Then*
(a) $t_1(P) \geq m - s + z - e$,
(b) $t_0(Q) \geq m - d + z - e$,
(c) $t_1(I) \leq e$.

Proof. (a) Since $(m; z)$-disjunct implies $(s; z+m-s)$-disjunct, a positive clone has at least $m - s + z$ elements not covered by the up-to-s inhibitors, hence resulting in at least $m - s + z$ positive outcomes, among which, at most e are changed to negative due to errors.

(b) Similar to (a).

(c) The only possibility that I appears in a positive test is due to error. □

Define

$$O = \{C \in N \mid t_1(C) \leq e\},$$

where N is the set of all clones. For $m+z \geq s+2e+1$, by Lemma 8.2.4, $t_1(P) \geq e+1$. Hence, O contains all inhibitors but no positive clone. Let S be a subset of O with $|S| = s$. Define $t_0^S(C)$ and $t_1^S(C)$ as before, except a negative test is changed to positive if it intersect a column in S. Further, define

$$t_0^*(C) = \min_{S \subseteq O} t_0^S(C).$$

Lemma 8.2.5 *Suppose M is $(m; z)$-disjunct with $m \geq d + s$. Then*
(a) $t_0^*(P) \leq e$,
(b) $t_0^*(Q) \geq m - d - s + z - e$,
(c) $t_0^*(I) \geq m - s + z - e$.

Proof. (a) There exists an $S \subseteq O$ which contains the inhibitor-set. For that S, $t_0^S(P) \leq e$ since error is the only reason for P being a negative test.

(b) Since $(m; z)$-disjunct implies $(d+s; m-d-s+z)$-disjunct, a negative clone has at least $m-d-s+z$ elements not covered by the up-to-d positive clones and the s columns in S, resulting in that many negative tests, among which at most e can be changed to positive due to errors.

(c) Similar to (b) except an inbitor does not have to be concerned for elements covered by positive clones. □

Proof of Theorem 8.2.3. We compute $t_0^*(C)$ for every C not in O. Substituting $m = d+s$ and $z = 2e+1$, then

$$t_0^*(P) \leq e,$$

and

$$t_0^*(I) \geq t_0^*(Q) \geq e+1.$$

Hence $\{C \in N \setminus O \mid t_0^*(C) \leq e\}$ is the set of positive clones. □

To check whether $t_0^*(C) \leq e$ for every $C \in N \setminus O$, it takes

$$(n - |O|)\binom{|O|}{s} \leq (\frac{n}{s+1})^{s+1}$$

operations. The computation can be somewhat reduced by noting:

> Lemma 8.2.4(a) implies that C is negative if $t_1(C) < m - s + z + e = d + e + 1$, and
>
> Lemma 8.2.4(b) implies that C is positive if $t_0(C) < m+d+z-e = s+e+1$.

Thus these clones can be excluded from $N \setminus O$ for further checking. Let N' denote the set after those exclusions. Hwang and Liu gave some special cases which N' is the empty set.

A popular way to construct d-disjunct matrix is through controlling the intersection number of two columns. Suppose this number is at most 1. Then $t_0(P) \leq s+e$ since each inhibitor column can cover at most one element of a given positive column. But as commented above, all columns C with $t_0(C) \leq s+e$ are identified as positive. So N' is an empty set and there is no need to compute $t_0^S(C)$.

Another special case is when M has constant column weight $d+s+2e+1$. Then for any column C, either $t_1(C) < d+e+1$ or $t_0(C) < s+e+1$. In either case C is identified, hence N' is the empty set.

Du and Hwang [8] showed that if the number of positive clones is known to be d, then the requirement on the disjunct matrix can be reduced. First, a lemma

Lemma 8.2.6 *Let M be a $(d+s; e+1)$-disjunct matrix. Then*

$$t_0^*(N) < t_0^*(P).$$

Proof. Suppose e_{01} errors occur which change a 0-outcome to 1, and e_{10} errors occur which change 1 to 0. Then

$$e_{01} + e_{10} \leq e.$$

When S covers the set of inhibitors, then

$$t_0^*(P) = t_0^S(P) = e_{10}.$$

On the other hand,

$$t_0^*(N) = \min_S t_0^S(N) \geq t_0(N) - s = (s + e + 1 - e_{01}) - s = e_{10} + 1.$$

□

Suppose $|\{C \in N \setminus O \mid t_0(C) < s + z + e\}| = p$.

Theorem 8.2.7 *Let M be a $(d+s; e+1)$-disjunct matrix. Then we can identify all positive clones in the $S(d,n)$ space by taking $d-p$ columns with the smallest $t_0^*(C)$ for $C \in N \setminus O$ together with the p positive clones already identified.*

Suppose we are also interested in identifying the inhibitor-set. For each P, let $\mathcal{S}(P)$ denote the collection of S such that $t_0^S(P) = 0$. Then the true inhibitor set is in $\cap_P \mathcal{S}(P)$. Unfortunately, we cannot conclude that $\cap_P \mathcal{S}(P)$ consists of a unique S since S, S' both in $\cap_P \mathcal{S}(P)$ simply means

$$(\cup S) \cap (\cup D) = (\cup S') \cap (\cup D),$$

which does not contradict the $(d+s)$-disjunctness of M.

8.3 Sequential and k-Round Algorithms

Farach *et al.* proposed a 3-step algorithm:

Step 1. Identify a positive pool S.
Step 2. Use S to identify all inhibitors and remove them.
Step 3. Identify all positives among the remaining clones.

They gave a probabilistic argument that a sequential algorithm with $O(s \log n)$ tests can be found in Step 1. In Step 2, similar to the explanation as given in the proof of Theorem 8.1.1, we can use a sequential group testing algorithm with $s \log n$ tests to identify all inhibitors. Once the inhibitors are all removed, then in Step 3, a sequential group testing algorithm with $d \log n$ tests identifies all positives. The total number of tests is $O((s+d)\log n)$, matching the lower bound in Theorem 8.1.2.

De Bonis and Vacarro [2] observed that Step 1 can be accomplished by an s-disjunct matrix. When the designated column is positive and the other s columns are inhibitors, then this row has a positive outcome and can serve as T. □

The minimum number of tests required for this modification is $O(s^2 \log n) + (d + s)\log n$.

Hwang and Liu [10] observed that the sequential algorithm of De Bonis and Vacano can be converted to a 3-round algorithm by using nonadaptive group testing algorithms in Steps 2 and 3 (the s-disjunct algorithm in Step 1 is already nonadaptive). Then the total number of tests is

$$O(s^2 \log n) + O(s^2 \log n) + O(d^2 \log n) = O((d^2 + s^2)\log n).$$

Hwang and Liu [10] also gave a 2-round algorithm with error tolerance.

The first round:
Pooling: Use a $(s; 2e+1)$-disjunct matrix.
Decoding: Output $O = \{C \in N \mid t_1(C) \le e\}$.

The second round:
Pooling: Use a $(d; 2e+1)$-disjunct matrix to test clones in the set $N \setminus O$.
Decoding: Output $D = \{C \in N \mid t_0(C) \le e\}$.

Lemma 8.3.1 *The set O contains all inhibitors and no positive.*

Proof. By Lemma 8.2.4(a), a positive clone should have at least $(e+1)$ 1-outcomes, hence no positive clone will appear in O. Besides, since all clones with at most e 1-outcomes are chosen into O, by Lemma 8.2.4(c), all inhibitors are contained in O. □

Theorem 8.3.2 *The set D contains all positives and nothing else.*

Proof. Since $N \setminus O$ contains no inhibitor, a $(d; 2e+1)$-disjunct matrix identifies up-to-d positive clones with e-error-correcting ability. □

De Bonis, Gasieniec and Vaccaro [1] gave a 4-stage algorithm:

Stage 1. Use a (s,d)-disjunct matrix to obtain a positive pool Q.

Stage 2. Use a (s,k)-disjunct matrix (with Q attached to every test) on $N \setminus Q$. Let S denote the set of columns not appearing in a positive test. Then $|S| < s + k$.

Stage 3. Use a (d,k')-disjunct matrix on $N \setminus S$. Let S' denote the set of clones not appearing in a negative test. Then $|S'| < d + k'$.

Stage 4. Individually checking every clone in $S \cup S'$ to obtain the set D of positive clones.

Here, k and k' are algorithm variables to be set.

Theorem 8.3.3 $D \subseteq S \cup S'$.

Proof. In Stage 2, a test is negative if and only if it contains an inhibitor. Since the matrix is (s, k)-disjunct, S contains all inhibitors. Thus $N \setminus S$ contains only positive and negative clones. So, a (d, k')-disjunct matrix outputs the set S' containing all positive clones in $N \setminus S$. But S may still contain positive clones and negative clones. Stage 4 identifies all positive clones in $S \cup S'$. □

Theorem 8.3.4 *The 4-stage algorithm requires $O(\max\{d, s\} \log n)$ tests, assuming d and s are much smaller than n.*

Proof. The number of tests is upper bounded by

$$t(s, d, n) + t(s, k, n - |Q|) + t(d, k', n - |S|) + s + k + d + k'.$$

Select $k = s$ and $k' = d$, then the above upper bound is about

$$O(t(s, d, n)) + O(s \log n) + O(d \log n) + 2s + 2d. \tag{8.3.1}$$

In view of Lemma 2.6.5, i.e., for $s = O(d)$, $O(t(s, d, n)) = O(d \log n)$. Then (8.3.1) is reduced to $O(d \log n)$, matching the lower bound in Theorem 8.1.2.

8.4 Some Other Inhibitor Models

De Bonis and Vaccaro [3] proposed the k-inhibitor model under which the outcome of a test is positive if and only if it contains at least one positive and at most $k - 1$ inhibitors. They gave a 4-stage algorithm:

Stage 1. Find a positive pool.
Stage 2. Find a positive pool containing exactly $k - 1$ inhibitors.
Stage 3. Identify all inhibitors and remove them.
Stage 4. Identify all positives.

The reason it is divided into four stages is that each stage depends on the result from its immediate preceding stage. They gave the following implementation which holds when there are at most k inhibitors.

To do Stage 1, a new type of d-disjunct matrix needs to be introduced. A binary matrix is $(k\text{-out-of-}s, d)$-disjunct if for any $d + s$ columns $C_1, ..., C_{d+s}$, there exists a union of k columns selected from $C_1, ..., C_s$ which cannot cover the union of $C_{s+1}, ..., C_{s+d}$, i.e., there exists a row in which one of $C_{s+1}, ..., C_{s+d}$ appears but k out of $C_1, ..., C_s$ do not. Note that (s-out-of-s, d) is (s, d)-disjunct.

Let M be a $((s-k+1)$-out-of-$s, d)$-disjunct matrix. If $|D| = 0$, then all outcomes are negative and we detect it. If $|D| \geq 1$, then there must exist a row of M which contains a positive but some $s - k + 1$ inhibitors are absent. Hence by testing all pools in M, we identify this row as a positive pool.

To do stage 2, all columns are partitioned into three subsets: S consists of columns in the positive pool, R consists of columns removed from consideration and O consists of the other columns. At the beginning, S is the positive pool obtained in Stage 1 and R is empty. Let H be a half of O or barely over half if $|O|$ is odd. Test $S \cup H$. If negative, assign $O \setminus H$ to R and S remains invariant (hence $O = H$). If positive, set $S = S \cup H$ and R remains invariant (hence $O = O \setminus H$). In any case, O is reduced about half in each iteration. Note that S is always presented as a positive pool, while $R \cup O$ always contains at least k inhibitors. Hence when O is a singleton, S must contain exactly $k - 1$ inhibitors.

To do Stage 3, we use S obtained in Stage 2. Let $\bar{S} = R \cup O$ and $S' \subseteq \bar{S}$. Then the outcome of testing $S \cup S'$ is positive if and only if S' contains no inhibitor. Therefore we can use a group testing algorithm to identify all inhibitors in $\bar{S}$, by attaching S to every pool.

Finally, after all inhibitors in $\bar{S}$ are removed, there are only $d-1$ inhibitors (those in S) left. So we can use a group testing algorithm to identify all positives in Stage 4.

De Bonis and Vaccaro gave an upper bound and a lower bound of $t(k$-out-of-$s, d, n)$.

Theorem 8.4.1

$$t(k\text{-}out\text{-}of\text{-}s, d, n) = \begin{cases} O((s+d)\log n/(s+d)) & for\ 20 < (s-k)/d, \\ O(s(s+d)/d(s-k+1)\log n/(s+d)) & for\ 20 \geq (s-k)/d. \end{cases}$$

Theorem 8.4.2

$$t(k\text{-}out\text{-}of\text{-}s, d, n) \geq \frac{1}{1+\log e} \lfloor \frac{s}{d(s-k+1)} \rfloor \log \frac{n-d+1}{sd}.$$

By imitating the Hwang-Sos construction of d-disjunct matrix, they also gave a construction of (k-out-of-s, d)-disjunct matrix with at most $24(s/(s-k+1))^2 \log n$ tests. Further, they showed that Stage 2 requires $\log n$ tests, Stage 3 $s \log n$ tests and Stage 4 $d \log n$ tests. Therefore the total number of tests is $O(\{[s/(s-k+1)]^2 + s + d\} \log n)$.

If the actual number of inhibitors is less than $s - k + 1$, then all tests in Stage 2 must have positive outcomes. When this all-positive phenomena is obtained, we detect the fact that the number of inhibitors among the n clones is less than $s-k+1$. Hence the existence of inhibitors is insequential. We can skip Stage 3 and go directly to Stage 4.

Chang and Hwang [4] gave a 1-round pooling design with a new type of disjunct matrices as follows: A binary matrix is called a (d + k-out-of-s)-disjunct matrix if for any $d+s+1$ columns $C_0, C_1, ..., C_{d+s}$, there are k columns among $C_{d+1}, ..., C_{d+s}$ such that the union of them together with $C_1, ..., C_d$ cannot cover C_0, in other words, there always exists a row intersecting C_0 but none of $C_1, ..., C_d$ and not intersecting at least k of $C_{d+1}, ...C_{d+s}$. Clearly, (d + 0-out-of-s)-disjunct is simply d-disjunct and (d + s-out-of-s)-disjunct is $(d+s)$-disjunct. $(d+k)$-disjunct implies (d + k-out-of-s)-disjunct for all $s \geq k$. Similarly, we define (d + k-out-of-s; z)-disjunct if in the above defintion "there always exists a row" is changed to "there always exist z rows".

Theorem 8.4.3 *A $(d + (s-k+1)$-out-of-$s; 2e+1)$-disjunct matrix can identify all positive clones under the $(\bar{d}, \bar{s})$ k-inhibitor model with at most e errors.*

Proof. For a given set S, denote by $t^S(C)$ the number of negative outcomes of C except a negative outcome is changed to positive if that test intersects S at least k times. With the change of definition of $t_0^S(C)$, the decoding procedure described in the proof of Theorem 8.2.3 is also applicable here.

Note that

$$t_o^*(P) = t_0^{S^*}(P) \leq e$$

where S^* contains the set of I and

$$t_0^*(I) \geq t_0^*(Q) \geq e+1$$

since there exist $e+1$ rows intersecting Q and at most $k-1$ Is, but none of D (hence with negative outcome). □

Corollary 8.4.4 *Suppose a $(d+s-k+1; 2e+1)$-disjunct matrix is used. Then R can be taken from all $(s-k+1)$-subsets of the column sets.*

The procedure in Corollary 8.4.4 reduces the number of S to check, as compared to the general model by reducing the upper bound of $|S|$ from s to $s-k+1$.

In general, one can consider a $x \times y$ inhibitor model in which every x inhibitors cancels the positive effect of every y positive clones. Besides the mathematical complexity of dealing with all these variation models, there is also the practical questions of determining which model fits the reality. Fortunately, Chang and Hwang observed that the decoding procedure and analysis given in Theorem 8.2.3 for the 1-inhibitor model actually works for all $x \times y$ inhibitor models if S is chosen from N, not just from O, and C from N, not just from N/O. Therefore there is no need to pinpoint the actual model fitting the reality. Of course, $\binom{N}{s}$ is a much larger set than $\binom{O}{s}$, this is the price we pay for not knowing the exact model.

We now state the general result.

Theorem 8.4.5 *A $(d+s; 2e+1)$-disjunct matrix identifies all positive clones in a $(\bar{d}, \bar{s})$ $x \times y$ inhibitor model with at most e errors.*

In particular, $\{C \in N \mid t_0^*(C) \leq e\}$ is the set of positive clones.

Suppose in the $(\bar{d}, \bar{r})$-complex model there are at most s inhibitor complexes, we call it the $(\bar{d}, \bar{r}, \bar{s})$-inhibitor complex model. $\bar{e}$ can be added to the set of parameters if there are at most e errors. Chang, Chang and Hwang [5] extended Theorem 8.2.3 to the complex model.

Theorem 8.4.6 *A* $(d+s, r; 2e+1]$*-disjunct matrix can identify all positive complexes in the* $(\bar{d}, \bar{r}, \bar{s}, \bar{e})$*-inhibitor complex model.*

Proof. Let X denote a complex. Define $t_0(X)$ (or $t_1(X)$) to be the number of rows containing X and having a negative (or positive) outcome. Set

$$O = \{X \in \binom{N}{s} \mid t_1(X) \leq e\}.$$

Then O contains all inhibitor complexes since an inhibitor complex must have a negative outcome except when an error occurs.

Let O^* denote the set of columns in O, and $S \subseteq O^*$ a set of s columns. For a positive complex P,

$$t_0^*(P) = \min t_0^S(P) \leq e,$$

which occurs when S intersects each of the up to s inhibitor complexes. On the other hand, for a negative complex Q,

$$t_0^*(Q) \geq e+1,$$

since by the $(d+s, r; 2e+1]$-disjunct property, there exist at least $2e+1$ rows covering Q but not a set of $d+r$ columns while the set intersects each of the up to d positive complexes and up to s inhibitor complexes. In other words, no inhibitor or positive complex appears in these rows, hence they must have negative outcomes. Since at most e of these negative outcomes can be changed to positive due to errors, $t_0^*(Q) \geq e+1$.

Thus we can separate positive and negative complexes by checking $t*_0(X)$ for all $X \in N \setminus O$.

Further, Chang, Chang and Hwang extended Theorems 8.4.3 and 8.4.5 to their complex version. Here we only state the latter extension. Theorem 8.4.6 says that $(d+s, r; 2e+1)$-disjunct matrix can identify all positive complexes in the $(\bar{d}, \bar{r}, \bar{s}, \bar{e})$ general inhibitor complex model. In particular, $\{X \in H \mid t_0^*(X) \leq e\}$ is the set of positive complexes.

References

[1] A. De Bonis, L. Gasieniec and U. Vaccaro, Optimal two-stage algorithms for group testing problems, *SIAM J. Comput.*, 34 (2005) 1253-1270.

[2] A. De Bonis and U. Vaccaro, Improved algorithms for group testing with inhibitors, *Inform. Proc. Lett.*, 67(1998) 57-64.

[3] A. De Bonis and U. Vaccaro, Constructions of generalized superimposed codes with applications to group testing and conflict resolution in multiple access channels, *Theor. Comput. Sci. A.*, 306 (2003) 223-243.

[4] F. H. Chang and F. K. Hwang, The identification of positive clones in a general inhibitor model, preprint, 2005.

[5] F. H. Chang, H. L. Chang and F. K. Hwang, Pooling designs for clone library screening in the inhibitor-complex model, preprint, 2006.

[6] A. G. Dyachkov, A. J. Macula, D. C. Torney and P.A. Villenkin, Two models of nonadaptive group testing for designing screening experiments, in A. C Atkinson, P. Hackl and W. G. Muller (Eds.): Proc. 6th Intern. Workshop on Model-oriented design and analysis, pp. 63-75, Physica-Velay, 2001.

[7] A. Dyachkov and V. Rykov, Bounds on the length of disjunctive code, *Prob. Control Inform. Thy.*, 11 (1982) 7-13.

[8] D.-Z. Du and F. K. Hwang, Identifying d Positive Clones in the Presence of Inhibitors, *International Journal of Bioinfomatics and Applications*, 1 (2005) 162-168.

[9] M. Farach, S. Kannan, E. Knill and S. Muthukrishuan, Group testing problems with sequences in experimental molecular biology, Proc. Compression and Complexity of Sequences, B. Cayertieu *et al.* (Eds.) IEEE Press, 1997, pp. 357-367.

[10] F. K. Hwang and Y. C. Liu, Error-Tolerant pooling designs with inhibitors, *J. Comput. Biol.*, to appear.

9
Hyperplane Designs

For convenience or practicality, clones are often stored in 2-dimensional arrays in a clone library and a test consists of a row or a column in the array. Since the set of tests consisting of all rows and all columns does not constitute a separating matrix, positive clones can be unresolved. To improve on the performance, one can use many arrays, add tests other than rows and columns, or increase the dimension of the array.

9.1 Introduction

It is popular to store clones in a 2-dimensional array since the storage is compact and orderly. For example, Green and Olson [12] reported a yeast-artificial-chromosome (YAC) library whose 23,000 clones are stored in 96-well microtiter plates. The clones are grown on nylon filters in a 16×24 grid of 384 colonies following inoculation from four microtiter plates. So the 23,000 clones are spread into 60 such filters.

A hierarchical scheme to identify all positive clones in the library is to work on each plate, or each colony, independently. In either case, the problem becomes to identify all positive clones in an $r \times c$ grid. It is convenient to take advantage of this set-up by having all the rows and columns as pools. The main advantage is the extreme easiness in collecting the pools, thus minimizing the chance of errors. Of course, a pooling design constituted in this way may not be optimal. By viewing an array as a plane and its rows and columns as the 1-dimensional subspaces such an array is a hyperplane design.

In general, when a transversal design in the form of a set of identical hyperplanes and the pools are subspaces of a given dimension, it is called a *hyperplane design*.

Note that a positive clone will yield a positive outcome for all subspaces containing it. Hence the clones at intersection of positive subspaces are prime suspect for positive clones. But, they do not have to be all positive if more than one positive clones exist. A counterexample had been given in Section 3.4. To solve this problem, multiple arrays were studied in Section 3.4 with two conditions. To test d positive clones, $d+1$ 2-dimensional arrays satisfying the unique collinearity condition are required.

In general, because perhaps there do not exist enough number of arrays satisfying the unique collinearity condition, or it would be too costly to do that many tests, or experimental errors may exist, we always assume that the testing on the array provides a candidate set of positive clones which will be individually tested for confirmation in the second stage. The goal is to minimize the total number of tests in both stages.

When the number of clones is much larger than a practical array size rc, there are three ways to proceed:

1. Partition the n clones into many subsets and for each subset, apply the above array construction.

2. The array construction is applied to the whole n clones. Namely, each array may involve a different subset of rc clones, but together they satisfy the unique collinearity condition. We call such an array an *incomplete array.*

3. Use an m-dimensional array, $m > 2$, to store the n clones.

These three approaches lead to three different graph-theoretical problems. Let Q_{rc} denote an $r \times c$ array, which can be seen as the graph $K_r \times K_c$ with all cells as vertices and an edge exists between two cells if and only if they are in the same row or the same column. Denote $Q_q = Q_{qq}$ and Q_q^m the m-dimensional $q \times \cdots \times q$ array. In the first approach, suppose the ith subset has n_i clones with $\sum n_i = n$. Theorem 3.4.2 shows that necessarily $n_i = q_i^2$ for some q_i. Then the problem is to pack K_n into Q_{q_i}. In the second approach, the problem is to pack K_n into Q_{rc}, while in the third approach, the problem is to pack K_n into Q_q^m with $n = q^m$.

Mathematically, the first problem is a special case of the second problem, and also a special case of the third problem. Section 7.2 will deal with packing K_{q^m} into Q_q^m, Sections 7.3 and 7.4 packing K_n into Q_{rc}.

Hwang [15] proved

Theorem 9.1.1 *Necessary and sufficient conditions for the existence of $g \geq 2$ grids satisfying the unique collinear condition are*

(i) $r = c$,

(ii) there exist $2(g-1)$ orthogonal Latin squares of order r.

Proof. Note that the difference between Theorem 3.4.2 and Theorem 9.1.1 is on condition (i). The sufficiency of Theorem 9.1.1 is the same as the sufficiency of Theorem 3.4.2. We need only to prove the necessity of condition (i). Let $G_0, G_1, ..., G_{g-1}$ denote the g grids. Suppose to the contrary that $r \neq c$, say $r > c$. By the pigeonhole principal, a column of G_1 must contain at least two elements from the same row of G_1, contradicting the unique collinear condition. □

We give some properties of Q_q^m when lines are tests.

Theorem 9.1.2 *Q_q^m has mq^{m-1} tests, is $\overline{(2^{m-1}-1)}$-separable and $(m-1)$-disjunct.*

Proof. The number of lines orthogonal to a given dimension is q^{m-1}. Since a vertex v has m lines while any other vertex can share at most one line with v, Q_q^m is d-disjunct with $d = (m-1)/\lambda = m-1$. □

With k-dimensional plane as tests, it is well-known that Q_q^m has $\binom{m}{k} q^{m-k}$ tests.

9.2 m-Dimensional Arrays

El-Zanati *et al.* [7] proved

Theorem 9.2.1 *Suppose that q is a prime power. Write $(q^m-1)/(q-1) = um+v$, $0 \le v < m$. Then K_{q^m} can be decomposed into u copies of Q_q^m and q^{d-v} vertex-disjoint copies of Q_q^v.*

A set of arrays is *compatible* if any two clones x, y collinear in a given array are not collinear in the other array. Since each clone is collinear with $m(q-1)$ other clones in a single array, while the total numbers of clones other than the given clone is $q^m - 1$, at most $u = \lfloor (q^m-1)/(m(q-1)) \rfloor$ compatible arrays can exist. If a set of compatible arrays attends this upper bound, we refer to this set as a *full* set of compatible arrays. For $d \ge 2$, Theorem 9.2.1 says that if q is a prime power, there is a full set of compatible arrays. The assumption of prime power is necessary. For example, if $q = 6$, then no two arrays are compatible. We now introduce a simple method to construct a full set of compatible array for q a prime power.

For easier implementation, De Jong *et al.* [6] and Barillot *et al.* [1] proposed using transforming matrices to construct compatible arrays.

Consider an m-dimensional $q \times \cdots \times q$ array G. The coordinates of each cell form an m-dimensional vector $(x_1, .., x_m)$. Let A be an $m \times m$ matrix. Then AG is the m-dimensional array where the entry of cell $(x_1, ..., x_m)^T$ in G is sent to cell $A(x_1, .., x_m)^T$ in AG. A is often called a *transforming matrix.*

How can A make $\{A(x_1, ..., x_d)^T \mid 0 \le x_1, ..., x_d \le q-1\} = \{(x_1, ..., x_d)^T \mid 0 \le x_1, ..., x_d \le q-1\}$? A simple way is to consider $x_1, ..., x_d$ as elements in the finite field of order q so that all coordinate vectors form an m-dimensional vector space over the field $GF(q)$. In this case, an $m \times m$ matrix A is a transforming matrix if and only if it transforms the whole space to the whole space, i.e., it is nonsingular. Note that the finite field $GF(q)$ exists if and only if q is a prime power. Therefore, we assume that q is a prime power throughout this section.

Let I_d be the $d \times d$ identity matrix. Then $\{I_d, A\}$ for a transforming matrix A is *efficient* if I_dG and AG are compatible for any G. In general, a set of transforming matrices $\{A_1, ..., A_k\}$ is an *efficient set* if $\{A_1G, ..., A_kG\}$ are compatible for any m-dimensional array G. An efficient set is *full* if the induced compatible set of arrays is full. Huang, Hwang and Ma [13] (also see [9]) gave a necessary and sufficient condition for an efficient set, and also a simple construction of a full set of transforming matrices when q is a prime power.

Lemma 9.2.2 *Let $A_1, ..., A_k$ be nonsingular matrices. Then $\{A_1, ..., A_k\}$ is an efficient set if and only if no two column vectors in matrices $A_1^{-1}, ..., A_k^{-1}$ are parallel.*

Proof. Let G be a d-dimensional array of order q. Let A and A' be two nonsingular matrices of order d over finite field $GF(q)$. Let $V_1, ..., V_d$ be column vectors of A^{-1} and $V'_1, ..., V'_d$ column vectors of $(A')^{-1}$. As a transforming matrix, A transforms the entry in cell $x = (x_1, ..., x_d)^T$ of G to cell $x^A = (x_1^A, ..., x_d^A)^T$ of the new array AG if $x^A = Ax$, i.e., $A^{-1}x^A = x$. Two clones x and y are collinear in AG if and only if the new coordinates x^A and y^A of x and y are different in exactly one component. Therefore,

$$x - y = A^{-1}(x^A - y^A) = \alpha V_i$$

for some $\alpha \in GF(q)$ and some column V_i of A^{-1}.

Similarly, two clones x and y are collinear in $A'G$ if and only if

$$x - y = \beta V'_j$$

for some $\beta \in GF(q)$ and some column V'_j of $(A')^{-1}$. Therefore, if two clones x and y are collinear in both AG and $A'G$, then $\alpha V_i = \beta V'_j$, i.e., V_i and V'_j are parallel.

Conversely, suppose A^{-1} has a column vector V_i parallel to a column vector V'_j of $(A')^{-1}$, i.e., $V_i = \alpha V'_j$ for some $\alpha \in GF(q)$. Choose a clone x in G and $y = x - V_i$. Then $x^A - y^A = A(x - y) = AV_i$ and $x^{A'} - y^{A'} = A'(x - y) = \alpha A'V'_j$. Note that both AV_i and $A'V'_j$ are vectors with exactly one nonzero component. Therefore, x^A and y^A are collinear and $x^{A'}$ and $y^{A'}$ are collinear. □

Lemma 9.2.3 *Let $A_1, ..., A_k$ be nonsingular matrices. Suppose $A_iA_j = A_jA_i$ for $1 \le i < j \le k$. Then $\{A_1, ..., A_k\}$ is an efficient set if and only if no two column vectors in matrices $A_1, ..., A_k$ are parallel.*

Proof. Note that there are two parallel columns in A_i^{-1} and A_j^{-1} if and only if there are two parallel columns in $(A_iA_j)A_i^{-1}$ and $(A_iA_j)A_j^{-1}$. Since $A_iA_j = A_jA_i$, we have $A_j = (A_iA_j)A_i^{-1}$ and $A_i = (A_iA_j)A_j^{-1}$. Therefore, Lemma 9.2.3 follows from Lemma 9.2.2. □

Example 9.2.1.

$$I_2 = \begin{pmatrix} 1 & 0 \\ 0 & 1 \end{pmatrix}, A_1 = \begin{pmatrix} 1 & 1 \\ 1 & 2 \end{pmatrix}, A_2 = \begin{pmatrix} 1 & 1 \\ 3 & 4 \end{pmatrix}$$

is a full set of transforming matrices and generates the following compatible arrays over $GF(5)$:

$$G_0 = \begin{pmatrix} (0,0) & (0,1) & (0,2) & (0,3) & (0,4) \\ (1,0) & (1,1) & (1,2) & (1,3) & (1,4) \\ (2,0) & (2,1) & (2,2) & (2,3) & (2,4) \\ (3,0) & (3,1) & (3,2) & (3,3) & (3,4) \\ (4,0) & (4,1) & (4,2) & (4,3) & (4,4) \end{pmatrix}$$

$$A_1G = \begin{pmatrix} (0,0) & (4,1) & (3,2) & (2,3) & (1,4) \\ (0,4) & (1,0) & (0,1) & (4,2) & (3,3) \\ (4,3) & (3,4) & (2,0) & (1,1) & (0,2) \\ (1,2) & (0,3) & (4,4) & (3,0) & (2,1) \\ (3,1) & (2,2) & (1,3) & (0,4) & (4,0) \end{pmatrix}$$

$$A_2G = \begin{pmatrix} (0,0) & (4,1) & (3,2) & (2,3) & (1,4) \\ (4,2) & (3,3) & (2,4) & (1,0) & (0,1) \\ (3,4) & (2,0) & (1,1) & (0,2) & (4,3) \\ (2,1) & (1,2) & (0,3) & (4,4) & (3,0) \\ (1,3) & (0,4) & (4,0) & (3,1) & (2,2) \end{pmatrix}$$

For each array in this example, each cell contains a coordinate vector which is the coordinate vector of the clone in the first array.

Note that the nonexistence of an efficient set over $GF(q)$ does not imply the nonexistence of a compatible array sets. For example, there is no 2×2 efficient matrix over $GF(4)$. However, since there are 3 mutually orthogonal Latin squares of order 4, we can find a new array compatible with the standard array.

For any positive integer m, there exists an irreducible polynomial $p(x)$ of degree m over $GF(q)$, such that $\{x, x^2, ..., x^{q^m-2}, x^{q^m-1} = 1\}$ modulo $p(x)$ is the set of all nonzero polynomials of degree less than m. In other words, the set of all non-zero polynomials of degree less than n forms a cyclic group with respect to polynomial multiplication modulo $p(x)$. The set of all scalars (polynomials of zero degree) is a cyclic subgroup of order $q-1$, of the form $x^k, x^{2k}, ..., x^{(q-1)k}$ where $k = \frac{q^n-1}{q-1}$. We call such $p(x)$ a *primitive polynomial.* The set of all polynomials over $GF(q)$ of degree less than m modulo a primitive polynomial is also of a finite order q^m.

Let $F = GF(q)$, $p(x) = p_0 + p_1x + \cdots p_{m-1}x^{m-1} + x^m$, $p_0 \neq 0$ be a primitive polynomial of degree m,

$$S = \begin{pmatrix} 0 & 0 & 0 & \cdots & 0 & -p_0 \\ 1 & 0 & 0 & \cdots & 0 & -p_1 \\ 0 & 1 & 0 & \cdots & 0 & -p_2 \\ \vdots & \vdots & \vdots & & \vdots & \vdots \\ 0 & 0 & 0 & \cdots & 0 & -p_{m-2} \\ 0 & 0 & 0 & \cdots & 1 & -p_{m-1} \end{pmatrix}$$

be the *companion matrix* of $p(x)$. It is easy to verify that S is a nonsingular matrix with determinant $-p_0$. Let

$$E_0 = (1, 0, 0, \cdots, 0)^T.$$

Then

$$SE_0 = (0, 1, 0, \cdots, 0)^T, ..., S^{m-1}E_0 = (0, 0, \cdots, 0, 1)^T.$$

Therefore,

$$\begin{aligned} \phi : a_0 + a_1x + \cdots + a_{m-1}x^{m-1} &\rightarrow \tau(a_0 + a_1x + \cdots + a_{m-1}x^{m-1})E_0 \\ &= a_0E_0 + a_1SE_0 + \cdots + a_{m-1}S^{m-1}E_0, \end{aligned}$$

is a linear-space isomorphism (preserving addition and scalar multiplication), which sends a polynomial with degree less than m to an m-dimensional column vector.

Lemma 9.2.4 *$\phi(x^i) = S^i E_0$ for any $i \geq 0$.*

Proof. For $0 \leq i \leq m-1$, $\phi(x^i) = S^i E_0$ by the definition of ϕ. To show it holds for $i \geq m$, it suffices to show that for any polynomial $g(x) = a_0 + a_1 x + \cdots + a_{m-1}x^{m-1}$, $\phi(xg(x)) = S\phi(g(x))$. Note that

$$\begin{aligned} xg(x) &= a_0 x + a_1 x^2 + \cdots + a_{m-1}x^m \\ &\equiv -a_{m-1}p_0 + (a_0 - a_{m-1}p_1)x + (a_1 - a_{m-1}p_2)x^2 + (a_{m-2} - a_{m-1}p_{m-1})x^{m-1} \\ &\quad (\text{mod } p(x)). \end{aligned}$$

Hence,

$$\begin{aligned} \phi(xg(x)) &= (-a_{m-1}p_0 + (a_0 - a_{m-1}p_1)S + (a_1 - a_{m-1}p_2)S^2 + (a_{m-2} \\ &\quad -a_{m-1}p_{m-1})S^{m-1})E_0 \\ &= S(a_0 + a_1 S + \cdots + a_{m-1}S^{m-1})E_0 - a_{m-1}\phi(p(x)) \\ &= S\phi(g(x)). \end{aligned}$$

□

Define $A_i = S^i$ for $0 \leq i \leq u-1$ where $u = \lfloor \frac{q^m-1}{m(q-1)} \rfloor$.

Theorem 9.2.5 *$\{A_0, A_1, ..., A_{u-1}\}$ is a full efficient set of transforming matrices.*

Proof. First, we note that $A_i A_j = A_j A_i$ for $0 \leq i < j \leq u-1$. Next, we note that column vectors of $A_0, A_1, ..., A_{u-1}$ are $E_0, SE_0, ..., S^{mu-1}E_0$. Suppose to the contrary that $S^i E_0 = \alpha S^j E_0$ for some $\alpha \in GF(q)$ and $0 \leq i \leq mu-1$. Since $\phi(x^i) = S^i E_0$, we have $\phi(x^i) = \alpha\phi(x^j)$. Thus, $x^i = \alpha x^j$. Hence $k \mid (j-i)$ where $k = (q^m-1)/(q-1)$. However,

$$j - i \leq mu - 1 = m \cdot \lfloor \frac{q^m - 1}{m(q-1)} \rfloor - 1 \leq k - 1,$$

a contradiction. □

Example 9.2.2. $p(x) = x^2 + x + 2$ is a primitive polynomial over $GF(3)$ with companion matrix

$$S = \begin{pmatrix} 0 & 1 \\ 1 & 2 \end{pmatrix}$$

$$\{x, x^2 = 2x+1, x^3 = 2x+2, x^4 = 2, x^5 = 2x, x^6 = x+2, x^7 = x+1, x^8 = 1\}$$

is the set of all nonzero elements of $GF(9)$. The corresponding 2-dimensional vectors are:

$$\phi(x) = S \cdot \begin{pmatrix} 1 \\ 0 \end{pmatrix} = \begin{pmatrix} 0 \\ 1 \end{pmatrix}, \phi(x^2) = S^2 \cdot \begin{pmatrix} 1 \\ 0 \end{pmatrix} = \begin{pmatrix} 1 \\ 2 \end{pmatrix},$$

$$\phi(x^3) = S^3 \cdot \begin{pmatrix} 1 \\ 0 \end{pmatrix} = \begin{pmatrix} 2 \\ 2 \end{pmatrix}, \phi(x^4) = S^4 \cdot \begin{pmatrix} 1 \\ 0 \end{pmatrix} = \begin{pmatrix} 2 \\ 0 \end{pmatrix},$$

$$\phi(x^5) = S^5 \cdot \begin{pmatrix} 1 \\ 0 \end{pmatrix} = \begin{pmatrix} 0 \\ 2 \end{pmatrix}, \phi(x^6) = S^6 \cdot \begin{pmatrix} 1 \\ 0 \end{pmatrix} = \begin{pmatrix} 2 \\ 1 \end{pmatrix},$$

$$\phi(x^7) = S^7 \cdot \begin{pmatrix} 1 \\ 0 \end{pmatrix} = \begin{pmatrix} 1 \\ 1 \end{pmatrix}, \phi(x^8) = S^8 \cdot \begin{pmatrix} 1 \\ 0 \end{pmatrix} = \begin{pmatrix} 1 \\ 0 \end{pmatrix}.$$

9.3 A $K_r \times K_c$ Decomposition of K_n

Let $D_{r\times c}(G)$ denote a decomposition of a graph G into $K_r \times K_c$. Hwang [15] gave necessary conditions for the existence of $D_{r\times c}(K_n)$.

Lemma 9.3.1 *Necessary condition for the existence of $D_{r\times c}(K_n)$ are*
(i) $(r + c - 2) \mid (n - 1)$, and
(ii) $rc(r + c - 2) \mid n(n - 1)$.

Proof. Each vertex of K_n has $n - 1$ neighbors while each vertex of a $K_r \times K_c$ has $r + c - 2$ neighbors in that array, which implies (i).

Also, there are $\binom{n}{2}$ pairs of neighbors in K_n while each $K_r \times K_c$ generates $rc(r + c - 2)/2$ pairs of neighbors. Condition (ii) follows immediately. □

Fu *et al.* [8] gave a general approach to construct a $D_{r\times c}(K_n)$. This general approach includes employing 2-dimensional difference sets to obtain cyclic constructions, and employing all sorts of combinatorial designs, like Steiner systems, pairwise balanced designs, resolvable designs and orthogonal Latin squares, to expand a construction.

Let the vertices of K_n be labeled by the numbers $0, 1, ..., n - 1$. Then the n labels generate $n-1$ distinct differences. A cyclic development of a $K_r \times K_c$ yields n $K_r \times K_c$ where the ith, $2 \le i \le n$ is obtained from the first by adding $(i - 1) \pmod{n}$ to every vertex label.

Lemma 9.3.2 *Suppose $rc(r+c-2)$ divides $n-1$ (implying n is odd). To decompose K_n into $K_r \times K_c$'s, it suffices to construct $(n - 1)/(rc(r + c - 2))$ starting $K_r \times K_c$ such that the difference are all distinct.*

Proof. By cyclically developing these $(n - 1)/(rc(r + c - 2))$ $K_r \times K_c$, we obtain $n(n-1)/(rc(r+c-2))$ $K_r \times K_c$ which constitute a decomposition of K_n. The reason is that two labels i and j will appear in a $K_r \times K_c$ developed from the starting $K_r \times K_c$ containing the differences $i - j$ and $j - i$. $\square$

Such a difference will be represented by a member of its residue class nearest to zero.

Example 9.3.1. For $n = 19, r = 2$ and $c = 3$, we have $(n-1)/(rc(r+c-2)) = 1$. The starting $K_r \times K_c$ is

$$\begin{array}{ccccc} 1 & - & 7 & - & 11 \\ | & & | & & | \\ 9 & - & 6 & - & 4 \end{array}$$

generates differences $\pm 6, \pm 9, \pm 4$ in the first row, differences $\pm 3, \pm 5, \pm 2$ in the second row, and difference $\pm 8, \pm 1, \pm 7$ in the three columns.

If $rc(r + c - 2)$ does not divide $n - 1$, then at least one starting $K_r \times K_c$ cannot be developed to complete a full cycle.

Example 9.3.2. For $n = 10, r = 2$ and $c = 3$, $(n - 1)/(rc(r + c - 2)) = 1/2$. We use the starting $K_2 \times K_3$ (the first one) but only for half a cycle.

$$\begin{array}{ccccc} 1 & - & 2 & - & 4 \\ | & & | & & | \\ 7 & - & 6 & - & 9 \end{array} \quad \begin{array}{ccccc} 2 & - & 3 & - & 5 \\ | & & | & & | \\ 8 & - & 7 & - & 0 \end{array} \quad \begin{array}{ccccc} 3 & - & 4 & - & 6 \\ | & & | & & | \\ 9 & - & 8 & - & 1 \end{array}$$

$$\begin{array}{ccccc} 4 & - & 5 & - & 7 \\ | & & | & & | \\ 0 & - & 9 & - & 2 \end{array} \quad \begin{array}{ccccc} 5 & - & 6 & - & 8 \\ | & & | & & | \\ 1 & - & 0 & - & 3 \end{array}$$

Note that the first row and the second row generate the same differences $\pm 1, \pm 3, \pm 2$. So the two rows in all five $K_2 \times K_3$ together constitute a full development of the set $\{1, 2, 4\}$. Similarly, the first column and the second column generates the same differences ± 4. Finally, the third column generates the difference ± 5. Since $5 \equiv -5 \pmod{10}$, a half cycle yields all pairs of difference 5.

Suppose that st vertices are partitioned into s subsets of t vertices each. Let $K_s(t)$ be the complete s-partite graph such that (i, j) is an edge if i and j are not in the same subset. It is useful to consider the decomposition of $K_s(t)$. Omit the subscript $r \times c$ in $D_{r \times c}(G)$ in the following. Similar to Lemma 9.3.1, we have

Lemma 9.3.3 *Necessary conditions for $D(K_s(t))$ to exist are*
(i) $(r + c - 2) \mid (s - 1)t$, *and*
(ii) $rc(r + c - 2) \mid (s - 1)st^2$.

The next result shows a connection between decomposing K_n and decomposing $K_s(t)$.

Theorem 9.3.4 *$D(K_{st+1})$ exists if $D(K_{t+1})$ and $D(K_s(t))$ exist.*

Proof. Partition $\{1, ..., st\}$ into s subsets $S_1, ..., S_s$ each having t vertices. Let $S*_i = S_i \cup \{0\}$. Let K_{S*_i} denote the complete graph on vertex set $S*_i$. Then $\cup_{i=1}^{s} D(K_{S*_i}) \cup D(K_s(t)) = D(K_{st+1})$. □

By naming the elements in S_i: $i, s+i, ..., (t-1)s+i$, for $i = 1, ..., s$, the differences which are multiples of s are taken care of in $D(K_{S*_i})$, $i = 1, ..., s$. There are still $(s-1)t$ out of the st differences left in K_{st+1}. Therefore, we have the following lemma.

Lemma 9.3.5 *Suppose $rc(r+c-2) \mid (s-1)t$. To obtain $D(K_s(t))$, it suffices to construct $(s-1)t/(rc(r+c-2))$ $K_r \times K_c$ with distinct differences which are not multiples of s.*

Example 9.3.3. For $s = 3, t = 9, r = 2$ and $c = 3$, $(s-1)t/(rc(r+c-2)) = 1$.

$$\begin{aligned} S_1 &= \{1, 4, 7, 10, 13, 16, 19, 22, 25\}, \\ S_2 &= \{2, 5, 8, 11, 14, 17, 20, 23, 26\}, \\ S_3 &= \{3, 6, 9, 12, 15, 18, 21, 24, 27\}. \end{aligned}$$

The starting $K_2 \times K_3$ is

$$\begin{array}{ccccc} 1 & - & 15 & - & 23 \\ | & & | & & | \\ 2 & - & 13 & - & 6 \end{array}$$

Again, when $rc(r+c-2)$ does not divide $(s-1)t$, some cycles are not fully developed.

$D(K_s(t))$ can be constructed recursively. Let V be a set of v elements, $\mathcal{B}$ be a collection of k-subsets of V and $\mathcal{G}$ be a partition of V into k classes, each of size n. A triple $(V, \mathcal{G}, \mathcal{B})$ is called a group divisible design with g groups of size v/g and block size k, denoted by $GD(v, g, k)$, if each pair of elements from V is either contained in exactly one group or is contained in exactly one block, but not both. Especially, in the case of $g = v$, a pair $(V, \mathcal{B})$ instead of $(V, \mathcal{G}, \mathcal{B})$ is called a Steiner 2-design with block size k and v elements, denoted by $S(2, k, v)$. Then the following theorem is obtained.

Theorem 9.3.6 *$D(K_g(nt))$ exists if $GD(v, g, s)$ and $D(K_s(t))$ exists. Especially, $D(K_v(t))$ exists if $S(2, s, v)$ and $D(K_s(t))$ exist.*

Proof. Replace each element in $GD(v,g,s)$ by a distinct t-part. Using the $D(K_s(t))$, all inter-part differences for parts in a block of $GD(v,g,s)$ are generated. What is missing is the inter-part differences among the v/g parts whose indices are in the same group of $GD(v,g,s)$, and the intra-part differences. Hence the construction yields $D(K_g(nt))$. □

Eexample 9.3.4. For $s = 3, t = 9, v = g = 7, r = 2$ and $c = 3$, partition $\{1, ..., 63\}$ into seven sets $S_1, ..., S_7$, each of nine elements. The seven blocks of $S(2,3,7)$ are $B_1 = \{S_1, S_2, S_4\}$, $B_2 = \{S_2, S_3, S_5\}$, $B_3 = \{S_3, S_4, S_6\}$, $B_4 = \{S_4, S_5, S_7\}$, $B_5 = \{S_5, S_6, S_1\}$, $B_6 = \{S_6, S_7, S_2\}$, $B_7 = \{S_7, S_1, S_2\}$, which is known as the Fano plane, that is, the projective plane of order 2.

Corollary 9.3.7 *$D(K_{vt+1})$ exists if $S(2,s,v)$, $D(K_{t+1})$ and $D(K_s(t))$ exist.*

Proof. By Theorem 9.3.6, $S(2,s,v)$ and $D(K_s(t))$ yield $D(K_v(t))$. To this design we add a point $\{0\}$. Apply $D(K_{t+1})$ to $G \cup \{0\}$, where G is a group of $D(K_v(t))$. □

Let $S(2, \{k_1, ..., k_z\}, v)$ denote a generalization of a Steiner 2-design where the block size can vary. Theorem 9.3.6 can be generalized to the following.

Theorem 9.3.8 *$D(K_v(t))$ exists if $S(2, \{s_1, ..., s_z\}, v)$ and $D(K_{s_i}(t))$, $i = 1, ..., z$, exist.*

Example 9.3.5. For $v = 12, s = 4, t = 9, r = 2$ and $c = 3$, we can construct $S(2, \{3,4\}, 12)$ from $S(2,4,13)$ by dropping the element 13. Since $D(K_3(9))$ and $D(K_4(9))$ both exist, $D(K_{12}(9))$ exists.

The following result gives a different construction.

Theorem 9.3.9 *$D(K_{(v-1)t+1})$ exists if $D(K_{(s-1)t+1})$, $S(2,s,v)$ and $D(K_s(t))$ exist.*

Proof. Delete the element v from $S(2,s,v)$. Then some blocks become of size $s-1$. Since v appears with any other element once, these blocks form a partition of $\{1, ..., v-1\}$. Let $\mathcal{B}_1$ denote the set of size-s blocks and $\mathcal{B}_2$ the set of size-$(s-1)$ blocks. Apply $D(K_{(s-1)t+1})$ to each $B_i' \cup \{0\}$ for $B_i' \in \mathcal{B}_2$, and apply $D(K_s(t))$ to each $B_i \in \mathcal{B}_1$. □

Example 9.3.6. For $v = 7, s = 3, t = 9, r = 2$ and $c = 3$, $(v-1)t+1 = 55$. Let $B_1, ..., B_7$ be the seven blocks of the Fano plane. After element 7 is deleted, $\mathcal{B}_1 = \{(S_1,S_2,S_4), (S_2,S_3,S_5), (S_3,S_4,S_6), (S_5,S_6,S_1)\}$, and $\mathcal{B}_2 = \{(S_4,S_5), (S_6,S_2), (S_1,S_3)\}$. Use $D(K_3(9))$ on each block of $\mathcal{B}$, and use $D(K_{19})$ on $S_4 \cup S_5 \cup \{0\}$, $S_6 \cup S_2 \cup \{0\}$ and $S_1 \cup S_3 \cup \{0\}$.

While Theorem 9.3.9 deals with deleting an element from $S(2, s, v)$, it is also possible to add an element. A Steiner 2-design $(V, \mathcal{B})$ is said to be *resolvable* with resolution u if $\mathcal{B}$ is partitioned into subcollections $\mathcal{R}_1, ..., \mathcal{R}_u$, called a resolution class such that each point in V is contained in $\mathcal{R}_i$ exactly once for any i.

Theorem 9.3.10 *$D(K_{(v+1)t+1})$ exists if a resolvable $S(2, s, v)$ system, $D(K_{t+1})$, $D_s(t)$ and $D(K_{s+1}(t))$ exist.*

Proof. Add a new element $v + 1$ to every block in a resolution class of the resolvable $S(2, s, v)$. Then some blocks become of size $s + 1$, which are handled by $D(K_{s+1}(t))$. □

Corollary 9.3.11 *$D(K_{(v+1)t+1})$ exists if a resolvable $S(2, s, v)$ system with resolution at least i, $D(K_{t+1})$, $D(K_s(t))$, $D_{s+1}(t)$ and $D(K_{it+1})$ exist.*

Proof. Let $(V, \mathcal{B})$ be a resolvable $S(2, s, v)$ with $u \geq i$ resolution classes $\{\mathcal{R}_1, ..., \mathcal{R}_u\}$. Add i new elements to V and let $\bar{V} = V \cup \{v + 1, v + 2, ..., v + i\}$. To each block B in $\mathcal{R}_j$ add an element $v + j$ and let $\bar{B} = B \cup \{v + j\}$ for $j = 1, 2, ..., i$. Make t copies of each element of $\bar{V}$ and let S_j be the set of t copies of each element j for $j = 1, 2, ..., v + i$. Moreover, add an element $\{0\}$. Apply $D(K_{s+1}(t))$ to each $\cup_{j \in \bar{B}} S_j$ for all $B \in \mathcal{R}_j$ and $j = 1, 2, ..., i$, and $D(K_s(t))$ on the elements $\cup_{j \in B} S_j$ for blocks B in $\mathcal{B} \setminus \cup_{j=1}^{i} \mathcal{R}_j$. In addition, apply $D(K_{t+1})$ to $S_j \cup \{0\}$ for $j = 1, 2, ..., v$ and $D(K_{it+1})$ on $(\cup_{j=1}^{i} S_{v+j}) \cup \{0\}$. □

In case of $r = c$, we may use "affine geometry" to obtain the decomposition.

Theorem 9.3.12 *For an even integer n and an odd prime power q, a $D_{q \times q}(K_{q^n})$ exists.*

Proof. Let α be a primitive element of $GF(q^n)$. Then each point of $GF(q^n)$ is represented by α^i. For convenience, let $\alpha^\infty = 0$. Here, for a prime power q, let $AG_i(n, q)$ be the set of i-dimensional subspaces and their cosets of $GF(q)$. We define a starting grid G_0 as follows:

$$\begin{array}{|ccccc|}
\hline
\alpha^\infty & \alpha^0 & \alpha^{2u} & \cdots & \alpha^{(2q-4)u} \\
\alpha^u & \alpha^0 + \alpha^u & \alpha^{2u} + \alpha^u & \cdots & \alpha^{(2q-4)u} + \alpha^u \\
\alpha^{3u} & \alpha^0 + \alpha^{3u} & \alpha^{2u} + \alpha^{3u} & \cdots & \alpha^{(2q-4)u} + \alpha^{3u} \\
\vdots & \vdots & \vdots & & \vdots \\
\alpha^{(2q-3)u} & \alpha^0 + \alpha^{(2q-3)u} & \alpha^{2u} + \alpha^{(2q-3)u} & \cdots & \alpha^{(2q-4)u} + \alpha^{(2q-3)u} \\
\hline
\end{array}$$

where

$$u = \frac{q^n - 1}{2(q - 1)}.$$

Thus, G_0 generates a $D_{q\times q}(K_{q^n})$ together with its cyclic shifts $\alpha^i G_0$ for $i = 0, 1, ..., u-1$ where each 2-flats gives a resolution class.

In fact, let α^i and α^j be two points in $AG(n, q)$. To count the number of rows and columns in the u grids containing α^i and α^j simultaneously, we have only to count the number of rows/columns such that 0 $(= \alpha^\infty)$ and $\alpha^i - \alpha^j$ occur together. We can represent $\alpha^i - \alpha^j = \alpha^l$ for some integer l. There is one line passing through the origin 0 and α^l, which proves the theorem. □

Example 9.3.7. Let $q = 3$ and $n = 2$. Then $\alpha, \alpha^2, ..., \alpha^8$ were given in Example 9.2.2. We give the exponent of α in G_0 and its cyclic shift G_1.

$$G_0 = \begin{pmatrix} \infty & 0 & 4 \\ 2 & 3 & 5 \\ 6 & 1 & 7 \end{pmatrix} \qquad G_1 = \begin{pmatrix} \infty & 1 & 5 \\ 3 & 4 & 6 \\ 7 & 2 & 0 \end{pmatrix}$$

A Steiner 2-design $(V, \mathcal{B})$ is said to be *cyclic* if there exists an automorphism σ of order $v = |V|$ which acts cyclically on V. For a cyclic Steiner 2 design $(V, \mathcal{B})$, the collection $\mathcal{B}$ of blocks is partitioned into orbits by σ. We choose block arbitrarily from each orbit and call it a starting block. Next, by combining starting blocks of a cyclic Steiner 2-design, we obtain the following theorem.

Theorem 9.3.13 *Let p be an odd prime power and $v \equiv p \pmod{2p(p-1)}$. If there exists a cyclic $S(2, p, v)$, then there exists a $D_{p\times p}(K_{pv})$.*

Proof. Let $t = (v - p)/2p(p - 1)$ and $u = v/p$. Firstly, we consider a cyclic $S(2, p, v)$. This design has $2t$ starting blocks with cycle length v and a single starting block with cycle length u. It is known that we can construct a $S(2, p, pv)$ from a $S(2, p, v)$ for a prime p (see Grannel and Griggs [11]). Let $a_x = (0, a_{x1}, ..., a_{x(p-1)})$ be starting blocks with cycle length v in $S(2, p, v)$ for $x = 0, 1, ..., 2t - 1$. According to the construction in [6, 11, 14], we can obtain some starting blocks of $S(2, p, pv)$ as follows for any $j = 0, 1, ..., p - 1$ and $x = 0, 1, ..., 2t - 1$.

$$(0, a_{x1} + jv, a_{x2} + 2jv, ..., a_{x(p-1)} + (p - 1)jv) \pmod{pv} \tag{9.3.1}$$

Making t pairs of starting blocks (a_x, a_y) in $S(2, p, v)$ by utilizing two starting blocks a_x and a_y, we obtain the following starting grid.

$$\begin{array}{|cccc|} \hline 0 & a_{x1} & \cdots & a_{x(p-1)} \\ a_{y1} & & \cdots & \\ \vdots & a_{xi} + a_{y_j} + (ij)v & \vdots & \\ a_{y(p-1)} & & \cdots & \\ \hline \end{array} \pmod{pv} \tag{9.3.2}$$

The rows and columns of the $G(p, p)$ contain each starting block of (9.3.1) exactly once for $j = 0, 1, ..., p - 1$. By checking the starting $G(p, p)$'s, we find all differences except for $\pm u, \pm 2u, ..., \pm(p^2 - 1)u/2 \pmod{pv}$.

By Theorem 9.3.12, it is known that a $D_{p\times p}(K_{p^2})$ exists. Multiplying each element of $D_{p\times p}(K_{p^2})$ by u, and by making its cyclic shift of length u, we obtain $D_{p\times p}(K_{pv})$ together with the starting $G(p,p)$'s (9.3.2). □

Using results given in this section, we can prove that the necessary condition given in Lemma 9.3.1 are also sufficient for the existence of $D_{r\times c}(K_v)$ with $(r,c) \in \{(2,2),(2,3),$
$(2,4),(3,3)\}$. However the (2, 2) and (2, 3) cases were first proved by other methods, the (2, 2) case as 4-cycle [16] and the (2, 3) case as a cubic graph [4]. The (2, 4) case was proved in [19] and the (3,3) case in [8].

Mutoh and Sarvate [20] extended the $K_r \times K_c$ grid to general dimensions, and give a direct decomposition of K_n into Q_2^3.

Let $P_{r\times c}(G)$ denote a packing of the graph G by $K_r \times K_c$, namely each edge of G appears in $P_{r\times c}(G)$ at most once. Such a packing is called a *resolvable packing* if the collection of $K_r \times K_c$ can be partitioned into classes, called resolution classes, such that each element appears exactly once in each resolution class. A resolvable $P_{r\times c}(G)$ is *maximal* if the number of resolution classes is maximal. Clearly, a resolvable $D_{r\times c}(G)$ is a maximal $P_{r\times c}(G)$. For $G = K_v$, a necessary condition for $P_{r\times c}(K_v)$ to exist is $\lfloor (v-1)/(r+c-2) \rfloor$. A resolvable packing of resolution u provides the choice of a resolvable packing of any resolution up to u to form a pooling design while vertex-symmetry is presented. For example, the following six array-blocks form a resolvable $P_{2\times 2}(K_8)$.

1	3	5	7	1	5	7	3	1	7	3	5
4	2	8	6	6	2	4	8	8	2	6	4

Mutoh, Jimbo and Fu [18] gave a construction of maximal resolvable $P_{2\times 2}(K_v)$ for all $v \equiv 0 \pmod 4$ to complement Theorem 9.3.12. They also gave a recursive construction of resolvable $P_{r\times c}(K_v)$.

The orthogonal array $OA(s,k,\lambda)$ of order s, degree k and index λ is an $s^2\lambda \times m$ matrix with entries from $S = \{0,1,...,s-1\}$ such that each $s^2\lambda \times 2$ submatrix contains every ordered pair of S exactly once.

Theorem 9.3.14 *Suppose $r \le c$. Then the existence of a resolvable $P_{r\times c}(K_v)$ with t resolution classes and an $OA(s,c+1,1)$ imply the existence of a resolvable $P_{r\times c}(K_{vs})$ with st resolution classes.*

Proof. Let $M = (m_{ij})$ denote the matrix $OA(s,c+1,1)$. By permuting the rows of $OA(s,c+1,1)$, if necessary, we may assume that for each column $j = 0,1,...,c-1$, the elements in the first s rows are all distinct, and so are the set of the second s rows, the third s rows, and so on. For each array $A^i = (a^i_{xy})$ in $P_{r\times c}(K_v)$, construct c arrays $A^{ii'} = ((a^i_{xy}, m_{i(x+y)}))$ for $0 \le i' \le s^2-1$, where $x+y$ is computed $\pmod c$.

We now show that $\{A^{ii'} \mid 0 \le i \le tv/rc-1, 0 \le i' \le s^2-1\}$ is a resolvable $P_{r\times c}(K_{vs})$ (note that $a^i_{xy} \in V = \{0,1,...,v-1\}$ and $m_{i'(x+y)} \in S$).

Since such pair (a_1, a_2) in V^2 appears at most once in A^i, and each pair (s_1, s_2) in S^2 appears exactly once in $OA(s, c+1, 1)$, (a_1, s_1) and (a_2, s_2) appear together at most once in any array in $\{A^{ii'}\}$. Hence $\{A^{ii'}\}$ is a packing. We now give the st resolution classes. For $u = 0, 1, ..., s-1$ and $w = 1, ..., t$, $\mathcal{R}_{uw}$ is a resolution, where

$$\mathcal{R}_{uw} = \{A^{ii'} \mid A^i \text{ is the } w\text{th resolution of } P_{r\times c}(K_v) \text{ and } i' = us, us+1, ..., us+s-1\},$$

since $(m_{us,j}, m_{us+1,j}, ..., m_{us+s-1,j}) = S$ for each $j = 0, 1, ..., c-1$. □

Using the same construction, Mutoh *et al.* proved

Theorem 9.3.15 *Suppose* $r \le c$. *Then the existence of a resolvable* $D_{r\times c}(K_m(n))$, *an* $OA(s, c+1, 1)$ *and a resolvable* $D_{r\times c}(K_{sn})$ *implies the existence of a resolvable* $D_{r\times c}(K_{smn})$.

Proof. Let $\mathcal{R}_{uw}$ be defined on $D_{r\times c}(K_m(n))$ and $OA(s, c+1, 1)$ similar to Theorem 9.3.14. Then we obtain $sn(m-1)/(r+c-2)$ resolutions of $D_{r\times c}(K_{smn})$. Let V_j denote the jth part of $K_m(n)$. By using the elements in $V_j \times S$ as vertices in $D_{r\times c}(K_{sn})$ for $j = 1, ..., m$, we obtain other $(sn-1)/(r+c-2)$ resolutions. They add up to $(smn-1)/(r+c-2)$ resolutions. □

By setting $K_m(n) = K_m$, i.e., each part has a single vertex, we have

Corollary 9.3.16 *The existence of a resolvable* $D_{r\times c}(K_m)$, *an* $OA(s, c+1, 1)$ *and a resolvable* $D_{r\times c}(K_{sn})$ *implies the existence of a resolvable* $D_{r\times c}(K_{sm})$.

A $P_{1\times c}(G)$ is of course a packing into ordinary blocks. Mutoh *et al.* showed how to obtain array-block packing from ordinary packing in the following two theorems.

Theorem 9.3.17 *Suppose* $r \le c$. *Then the existence of a* $P_{1\times c}(K_v)$ *with resolution* t *implies the existence of a* $P_{r\times c}(K_{rv})$ *of resolution* t.

Proof. We only give the array $A^i = (a^i_{xy})$, $1 \le i \le v(v-1)/c(c-1)$, and the resolution classes.

Let $B^i = (b^i_j)$ denote the ith block of $P_{1\times c}(K_v)$. Then $A^i = ((b^i_{x+y}, y))$, where $x+y$ are $\pmod c$.

$$R_w = \{A^i \mid B^i \text{ is a block in the } w\text{th resolution of } P_{1\times c}(K_v)\}$$

is a resolution class of $P_{r\times c}(K_{rv})$ for $w = 1, ..., t$. □

Theorem 9.3.18 *Suppose* $r \le c$. *Then the existence of a resolvable* $P_{1\times c}(K_v)$ *with resolution* t *and a resolvable* $P_{r\times c}(K_{rc})$ *with resolution* $s+1$ *imply the existence of a resolvable* $P_{r\times c}(K_{rv})$ *with resolution* $st+1$.

Proof. Again, we only give the construction and the resolution classes.

Let $M = (m^0_{xy})$ be one of the $s+1$ $r \times c$ array-blocks in $P_{r\times c}(K_{rv})$. Rename the elements in M^0 such that $m^0_{xy} = (xy)$. Then m^j_{xy} can be written as $(\alpha^j_{xy}, p^j_{xy})$, $0 \le \alpha^j_{xy} \le r-1$, $0 \le \beta^j_{xy} \le c-1$. For each block $B^i = (b^i_0, b^i_1, ..., b^i_{c-1})$ in $P_{1\times c}(K_v)$, and each M^j, $1 \le j \le s$, define the grid-block $A^{ij} = ((b^i_{\beta^j_{xy}}, \alpha^j_{xy}))$. Then we will show that

$$A^j_w = \{A^{ij} \mid B_i \text{ is in the } w\text{th resolution class of } P_{1\times c}(K_v)\}$$

is a resolution class. Since w has t choices and j has s choices, $\{A^j_w\}$ yields st resolution classes of $P_{r\times c}(K_{rv})$. One more is given by A^0_1, which can be viewed as

$$\begin{pmatrix} (0,0) & (1,0) & \cdots & (c-1,0) \\ (0,1) & (1,1) & \cdots & (c-1,1) \\ \vdots & \vdots & \ddots & \vdots \\ (0,r-1) & (1,r-1) & \cdots & (c-1,r-1) \end{pmatrix}$$

$$\vdots$$

$$\begin{pmatrix} (v-c,0) & (v-c+1,0) & \cdots & (v-1,0) \\ (v-c,1) & (v-c+1,1) & \cdots & (v-1,1) \\ \vdots & \vdots & \ddots & \vdots \\ (v-c,r-1) & (v-c+1,r-1) & \cdots & (v-1,r-1) \end{pmatrix}$$

Now consider two distinct elements (v,j) and (v',j') of $P_{r\times c}(K_{rv})$.

(i) $v = v'$. Then they appear together exactly once in A^0_1.

(ii) $v \ne v'$. If v and v' do not appear together in a block of $P_{1\times c}(K_v)$, then they don't appear together in any A^{ij} since the first coordinate of an element in A^{ij} is from B^i. If v and v' both appear in B^i, then (v,j) and (v',j') can be together only in one A^{ij} since the $\{M^j\}$ is a packing. □

For $r = c \ne 6$, there exists a pair of orthogonal Latin squares, or a $P_{r\times c}(K_{r^2})$ with resolution 2. Hence

Corollary 9.3.19 *For $r \ne 6$, the existence of a $P_{1\times r}(K_v)$ with resolution t implies the existence of a $P_{r\times r}(K_{rv})$ of resolution $t+1$.*

For $r = c$ an odd prime power, a $P_{r\times r}(K_{r^2})$ with resolution $(r-1)/2$ exists. Hence

Corollary 9.3.20 *For r an odd prime power, the existence of a $P_{1\times r}(K_v)$ with resolution t implies the existence of a $P_{r\times r}(K_{rv})$ of resolution $t(r-1)/2+1$.*

9.4 Efficiency

Berge *et al.* [2] defined *efficiency* as the expected number of positives identified per test. They advocated efficiency instead of number of tests as a measure of performance since efficiency can be compared among pooling designs with different numbers of positive clones. Along that line, it is more convenient to use the ratio $p = d/n$ instead of d as a design parameter for comparing pooling designs with different number of clones.

Let $f(q_1 \times q_2 \times \cdots \times q_m)$ denote the efficiency of an m-dimensional $q_1 \times q_2 \times \cdots \times q_m$ array. If $q_1 = q_2 = \cdots = q_m$, then we abbreviate it by $f(q^m)$. Phatarfox and Sudbury [23] proved

Theorem 9.4.1 $f(q_1 \times q_2) = p/\{q_1^{-1} + q_2^{-1} + p + (1-p)[1-(1-p)^{q_1-1}][1-(1-p)^{q_2-1}]\}$.

Proof. All positive clones are identified in an array design. The total number of tests is the $q_1 + q_2$ tests in the first stage plus the number of candidates. A clone becomes a candidate if and only if the row and the column incident to it are both positive. This happens either if the clone is positive or it is negative but the row and the column both contains a positive clone. The probability of this event for a given clone is

$$p + (1-p)[1-(1-p)^{q_1-1}][1-(1-p)^{q_2-1}].$$

It follows

$$\begin{aligned} f(q_1 \times q_2) &= \frac{q_1 q_2 p}{q_1 + q_2 + q_1 q_2 [p + (1-p)[1-(1-p)^{q_1-1}][1-(1-p)^{q_2-1}]} \\ &= \frac{p}{q_1^{-1} + q_2^{-1} + p + (1-p)[1-(1-p)^{q_1-1}][1-(1-p)^{q_2-1}]}. \end{aligned}$$

□

Corollary 9.4.2 $f(q^2) = p/\{2q^{-1} + p + (1-p)[1-(1-p)^{q-1}]^2\}$.

Theorem 9.4.1 can be readily extended to

Theorem 9.4.3 $f(q_1 \times q_2 \times \cdots \times q_m) = p/\{\sum_{i=1}^m q_i^{-1} + p + (1-p)\prod_{i=1}^m [1-(1-p)^{q_i-1}]\}$.

Corollary 9.4.4 $f(q^m) = p/\{mq^{-1} + p(1-p)[1-(1-p)^{q-1}]^m\}$.

Corollary 9.4.4 was first given by Berge *et al.* for $d = 3$, but their formula is slightly different from ours.

Let $f(q^m, k)$ denote the efficiency of an m-dimensional array with side q where each k-dimensional subarray is a pool. Then $f(q^m, 1) = f(q^m)$.

Theorem 9.4.5 $f(q^m, k) = p/\{\binom{m}{k} q^{-k} + p + (1-p)[1-(1-p)^{q^k-1}]^{\binom{m}{k}}\}$.

Proof. There are $\binom{m}{k}q^{m-k}$ k-dimensional hyperplanes, hence that many tests in the first stage. A clone becomes a candidate either it is positive or it is negative but the $\binom{m}{k}$ k-dimensional hyperplanes containing it all contain a positive clone. The probability of this event is

$$p + (1-p)[1-(1-p)^{q^k-1}]^{\binom{m}{k}}.$$

Hence,

$$\begin{aligned} f(q^m, k) &= \frac{q^m p}{\binom{m}{k}q^{m-k} + q^m\{p + (1-p)[1-(1-p)^{q^k-1}]^{\binom{m}{k}}} \\ &= \frac{p}{\binom{m}{k}q^{-k} + p + (1-p)[1-(1-p)^{q^k-1}]^{\binom{m}{k}}}. \end{aligned}$$

□

Finally, let $f(q^m, k, j)$ denote the efficiency of j copies of Q_q^m such that any two clones are in the same k-dimensional hyperplane at most once. Then using an argument analogous to the proof of Theorem 9.4.5, we obtain

Theorem 9.4.6 $f(q^m, k, j) = p/\{j\binom{m}{k}q^{-k} + p + (1-p)[1-(1-p)^{q^k-1}]^{j\binom{m}{k}}\}$.

In particular,

Corollary 9.4.7 $f(q^m, 1, j) = p/\{jdq^{-k} + p + (1-p)[1-(1-p)^{q-1}]^{jm}\}$.

Berge *et al.* also derived several upper bounds of efficiency.

Theorem 9.4.8 *The efficiency of any algorithm, sequential or nonadaptive, is upper bounded by* $p/[-p\log p - (1-p)\log(1-p)]$.

Proof. The denominator is the entropy function, representing the degree of uncertainty in bits. Since the test outcome is binary and hence can only provide one bit of information, the entropy function also represents the minimum number of tests needed. □

Recall that a 2-stage algorithm is called *trivial* if every candidate produced in the first stage must be tested in the second stage. Note that all array and transversal schemes discussed in this chapter are trivial algorithms.

Consider a 2-stage trivial algorithm T. Partition the clone into three categories:

Category 1. The K_1 clones not appearing in any stage-1 test.

Category 2. The K_2 clones tested individually in stage 1.

Category 3. The K_3 clones each appearing in at least one test of size ≥ 2 in stage 1.

Every category-1 clone has to be tested individually in stage 2. Let x be a category-3 clone, and let $t_s(x)$ be the number of stage-1 tests of size s containing x. Then the total number of stage-1 tests of size ≥ 2 is

$$t = \sum_{x=1}^{K_3} \sum_{s=2}^{K_3} t_s(x)/s$$

since each test of size s is counted s times. Let $E(T)$ denote the expected number of tests. Then

$$E(T) = K_1 + K_2 + \sum_{x=1}^{K_3} \sum_{s=2}^{K_3} t_s(x)/s + \sum_{x=1}^{K_3} P(x \in S_2),$$

where $x \in S_2$ is the event that x is tested in stage-2. Let x^+ and x^- denote the respective events that x is positive and x is negative. Then

$$\begin{aligned} P(x \in S_2) &= pP(x \in S_2 \mid x^+) + (1-p)P(x \in S_2 \mid x^-) \\ &= p + (1-p)P(x \in S_2 \mid x^-). \end{aligned}$$

Define $t(x) = \sum_{s=2}^{K_3} t_s(x)$. Let $T_i(x)$ denote the ith such test, and $T_i(x)$ denote the event that the outcome of $T_i(x)$ is positive. Then

$$\begin{aligned} P(x \in S_2 \mid x^-) &= P(T_1(x)^+ \mid x^-)P(T_2(x)^+ \mid T_1(x)^+, x^-) \\ &\quad \cdots P(T_t(x)^+ \mid T_1(x)^+, \ldots, T_{t(x)-1}(x)^+, x^-) \\ &\geq \prod_{i=1}^{t(x)} P(T_i(x)^+ \mid x^-) \\ &= \prod_{s=2}^{K_3} [1 - (1-p)^{s-1}]^{t_s(x)} \end{aligned}$$

the inequality is due to the fact that the events $T_1(x)^+, T_2(x)^+, ..., T_{t(x)}(x)^+$ are positively correlated. It follows that

$$\begin{aligned} E(T) &\geq K_1 + K_2 + \sum_{x=1}^{K_3} \{\sum_{s=2}^{K_3} t_s(x)/s + p + (1-p) \prod_{s=2}^{K_3} [1 - (1-p)^{s-1}]^{t_s(x)}\} \\ &\geq K_1 + K_2 + K_3 \min_{\{t_s\}} \{\sum_{s=2}^{K_3} t_s(x)/s + p + (1-p) \prod_{s=2}^{K_3} [1 - (1-p)^{s-1}]^{t_s(x)}\}, \end{aligned}$$

where the minimum is over all partitions of t into nonnegative integers $\{t_2, t_3, ..., t_{K_3}\}$.

Define $r_s = t_s/s$ and $R = \sum_{s=2}^{K_3} r_s$. Then r_s should be a nonnegative rational. This condition is relaxed with the effect that the minimum is further reduced and the inequality is not effected. Then we have

$$E(T) \geq K_1 + K_2 + K_3 \min_{\{r_s\}} \{R + p + (1-p) \prod_{s=2}^{K_3} [1 - (1-p)^{s-1}]^{sr_s}\}.$$

For a fixed p, $[1-(1-p)^{s-1}]^s$ has a unique minimum $s^* = 1+\log u/\log(1-p)$ where u satisfies the transcendental equation

$$-u\log u + (1-u)\log(1-u) = u\log(1-p).$$

For $p < 0.164$, there are two such positive values of u, the smaller one yields s^*. Then

$$\begin{aligned} E(T) &\geq K_1 + K_2 + K_3 \min_{\{r_s\}}\{R + p + (1-p)[1-(1-p)^{s^*-1}]^{s^*R}\} \\ &\geq K_1 + K_2 + K_3\{R + p + (1-p)[1-(1-p)^{s^*-1}]^{s^*R}\} \\ &= n - K_3\{1 - R^* - p - (1-p)[1-(1-p)^{s^*-1}]^{s^*R^*}\}, \end{aligned}$$

where $R^* = \log|(1-p)[1-(1-p)^{s^*-1}]^{s^*}|/\log|[1-(1-p)^{s^*-1}]^{s^*}|$ achieves the minimum. Thus we have

Theorem 9.4.9 *An upper bound $f(T)$ for any 2-stage trivial algorithm T is*

$$\frac{p}{1-\min_{K_3}(K_3/n)\{1-R^*-p-(1-p)[1-(1-p)^{s^*-1}]^{s^*R^*}\}}.$$

A positive clone can be confirmed in stage 1 if it appears in a test in which all other clones are confirmed to be negative, i.e., they all appear in some negative tests. If such a clone does not have to go through individual testing, the algorithm is called *intelligent.*

Corollary 9.4.10 *An upper bound of $f(T)$ for any 2-stage trivial algorithm T is*

$$\frac{p}{1-\min_{K_3}(K_3/n)\{1-R^*-(1-p)[1-(1-p)^{s^*-1}]^{s^*R^*}\}}.$$

Proof.

$$P(x\in S_2) \geq (1-p)P(x\in S_2 \mid x^-),$$

while the right-hand side represents the most favorable case that all positive clones are confirmed in stage 1. So the only difference between the upper bounds of the two classes of algorithms is that $P(x\in S_2)$ misses the term p. □

9.5 Other Transversal Designs

Suppose in a set of 2-dimensional arrays, one takes not only the rows and the columns as pools, but also the front diagonals. Then the set is called a *row-column-front* (RCF) design if each pair of clones appears in exactly one block. If furthermore, the back diagonals are also taken, then it is called a *union jack design* if each pair of clones appears in exactly one block.

Chateauneuf, Colbourn, Krehu, Lamken and Torney [5] first studied these designs and gave some sufficient conditions.

Theorem 9.5.1 *If* $n \equiv 5 \pmod 6$ *is a prime, then there exists an RCF design on* n *clones.*

Proof. Design for each slope $s \in GF(n) \cup \{\infty\}$ and each $b \in GF(n)$, the line $L_{sb} = \{(x, sx + b) \mid x \in GF(n)\}$, the $(n+1)n$ lines of L_{sb} are the blocks of the affine plane of order n and form a resolvable $(n^2, n, 2)$ Steiner design. The set $R_s = \{L_{sb} \mid b \in GF(n)\}$ forms a partition of the n^2 points (x, y).

For $n \equiv 5 \pmod 6$ a prime, an RCF design can be constructed from the affine plane. Partition $GF(n) \cup \{\infty\}$ into classes of size 3, $X_1, ..., X_{(n+1)/3}$ so that $X_1 = \{\infty, 0, 1\}$. Then the lines with slopes in X_1 form an array in which the parallel classes form rows, columns and front diagonals. Further, the group $PGL_2(n)$ acts 3-transitively on the set of slopes, hence there is a group element mapping X_1 to X_i for each $2 \leq i \leq (n+1)/3$. Applying this automorphism places the parallel classes corresponding to the slopes in X_i in the roles of rows, columns and front diagonals in an array. □

An RCF design of order 5 defined on $Z_5 \times Z_5$ is given here.

$$\begin{pmatrix} 00 & 10 & 20 & 30 & 40 \\ 01 & 11 & 21 & 31 & 41 \\ 02 & 12 & 22 & 32 & 42 \\ 03 & 13 & 23 & 33 & 43 \\ 04 & 14 & 24 & 34 & 44 \end{pmatrix} \begin{pmatrix} 00 & 12 & 24 & 31 & 43 \\ 34 & 41 & 03 & 31 & 22 \\ 13 & 20 & 32 & 44 & 01 \\ 42 & 04 & 11 & 23 & 30 \\ 21 & 33 & 40 & 02 & 14 \end{pmatrix}$$

The elements in the six blocks (0,0) are (10, 20, 30, 40), (01, 02, 03, 04), (11, 22, 33, 44), (12, 24, 31, 43), (34, 13, 42, 21), (41, 32, 23, 14). Note that each element other than (0,0) appears exactly once.

They also proved

Theorem 9.5.2 *If* $n \equiv 3 \pmod 4$ *is a prime, then there exists a union jack design of order* n*.*

We omit the proof since it is quite involved.

Let $TD(k, n)$ represents a transversal design with k parts. Chateauneuf *et al.* observed that a necessary condition for the RCF design is the containment of a cyclic $TD(3, n)$. This rules out the planes arising from the finite field and the translation planes as candidates to generate RCF designs.

Fu and Lin [10] gave alternative and simplied constructions derived from affine transformation for the designs in Theorems 9.6.1 and 9.6.2. They [9] further extended the construction for the latter to higher dimensions.

9.6 Two Recent Applications

A DNA chip was first proposed to sequence a target sequence by obtaining all k-subsets contained in it. It is well documented [21] that this approach leads to some serious difficulties. One of its disadvantages is that the DNA chip is not efficiently used since most of the k-mers in the chip do not hybridize. Hubbell [14] proposed the multiplex SBH in which many target sequences are pooled together in one application of the DNA chip. Namely, a k-mer in the chip will have a hybridization if one of the target sequences contains the dual of the k-mer as a subsequence. Naturally, the problem arises as to assigning which k-mers to which target sequence in the pool.

Hubbell proposed to use the 2-dimensional hyperplane design as the pooling designs for n target sequences. A row or a column in each array serves as a pool in one application of the DNA array. Let $K(y)$ denote the set of k-mers hybridized in the pool y. Suppose a target sequence s appears in p pools $P_1(s), P_2(s), ..., P_p(s)$ with hybridization. Then, the k-mers in s can be recovered from $\cap_{i=1}^{p} K(P_i(s))$. A k-mer not in s may be wrongly assigned to it if and only if all $P_i(s)$ contains it.

Suppose each target sequence is of length l. For $k << l$, a target sequence contains about l k-segments. We will ignore the dependence between adjacent k-segments and treat each k-segment as an independent k-subset. Further, suppose each pool contains q target sequences. Then a pool y contains about ql k-segments. Define $F = 4^k/ql$ and assume F large. Then $K(y)$ contains about ql (distinct) k-mers among which about $(q-1)l$ are not in T. The probability that among the k-mers assigned to s there is one not in s, i.e., the expected number of false positives, is about $4^k[(q-1)l/4^k]^p < 4^k/F^p$. Hence for $4^k << F^p$, the probability of no false positive is high. In other words $p > k \log_F 4$ should be enough to guarantee a high probability of no false positive in the set of k-mers assigned to s.

Under the above scenario, the total number of tests of sequencing $n = q^2$ target sequences is

$$pq \sim (k \log_F 4)\sqrt{n},$$

and the number of tests per target sequence is

$$k \log_F 4/\sqrt{n}.$$

Cai *et al.* [3] also proposed to use the 2-dimensional array design to pool target sequences for sequencing. Pool the target sequences in a row (or column) and consider it a library. Then each target sequence s appears in a unique pair of libraries, one row library $r(s)$ and one column library $c(s)$. Use shotgunning to obtain a set of contigs large enough such that each of $r(s)$ and $c(s)$ has a 4- to 5-fold coverage. Then the contigs in s will be sequenced with 8- to 10-fold coverage. Again, the problem of false assignment may occur, which can be analyzed in a similar way as done by Hubbell.

The main advantage of this set-up is to reduce the number of libraries from n to $2\sqrt{n}$. Since we do shotgun sequencing for every pair of row and column, we still perform a total of n shotgun sequencings, same as if we do the n target sequences individually. Alternatively, we can imitate Hubbell's method in doing a shotgun sequencing for each row and column, and then use their intersection of contigs in $r(s)$ and $c(s)$ to determine the set of contigs in s, thus reducing the number of shotgun sequencing to $2\sqrt{n}$.

References

[1] E. Barillot, B. Lacroix and D. Cohen, Theoretical analysis of library screening using a N-dimensional pooling strategy, *Nucleic Acids Res.*, 19 (1991) 6241-6247.

[2] T. Berger, J. W. Mandell and P. Subrahmanya, Maximally efficient two-stage screening, *Biometrics*, 56 (2000) 833-840.

[3] W. W. Cai, C. W. Chow, S. Damani, G. Simon and A. Bradly, A SSLD anchored BAC framework map of the mouse genome, *Nat. Genet.*, to appear.

[4] J. E. Carter, Designs on cubic multigraphs, Ph.D. Thesis, McMaster University, 1989.

[5] M. A. Chateauneuf, C.C. Colbourn, D.L. Kreber, E.R. Lamken and D.C. Torney, Pooling, lattice square and union jack designs, *Ann. Combin.* 3 (1999) 27-35.

[6] P. J. De Jong, C. Aslanidis, J. Alleman and C. Chen, Genome mapping and sequencing, Proc. Cold Spring Harbor Workshop, New York 1990, pp. 48.

[7] S. El-Zanati, M. Plantholt and C. V. Eynden, Graph decompositions into generalized cubes, *Ars Combinatoria*, 49 (1998) 237-247.

[8] H. L. Fu, F. K. Hwang, M. Jimbo, Y. Mutoh and C. L. Shiue, Decomposing complete graphs into $K_r \times K_c$'s, *J. Statist. Plan. Infor.*, to appear.

[9] T. S Fu and K. S. Lin, Lattice hypercube designs, *Disc. Appl. Math.*, 131 (2003) 673-680.

[10] T. S Fu and K. S. Lin, A uniformed construction of two-dimensional pooling designs, *Ann. Combin.* 8 (2004) 155-159.

[11] M. J. Grannel and T. S. Griggs, Product constructions for cyclic block designs: Steiner-2 designs, *J. Combin. Thy. A*, 42(1986) 129-183.

[12] E. D. Green and M. V. Olson, Systematic screening of yeast artificial chromosone libraries by use of the polymer chain reaction, *Proc. Nat. Acad. Sci., USA*, 87 (1990) 1213-1217.

[13] H. M. Huang, F. K. Hwang and J. F. Ma, Using transforming matrices to generate DNA clone grids, *Disc. Appl. Math.*, 129 (2003) 421-431.

[14] E. Hubbell, Multiplex sequencing by hybridization, *Journal of Computational Biology*, Vol 8, No 2 (2001) 141-149.

[15] F. K. Hwang, An isomorphic factorization of the complete graph, *J. Combin. Thy.*, 19 (1995) 333-337.

[16] A. Kotzig, On decomposition of complete graph into $4k$-gons, *Math.-Fyz. Cas.*, 15 (1965) 227-233.

[17] R. J. McEliece, *Finite Fields for Computer Scientists and Engineers*, Kluwer Berlin, 1987.

[18] Y. Mutoh, M. Jimbo and H. L. Fu, A resolvable $r \times c$ grid-block packing and its application to DNA library screening, preprint.

[19] Y. Mutoh, T. Monhara, M. Jimbo and H. L. Fu, The existence of 2×4 grid designs and its applications, *SIAM J. Disc. Math.*, to appear.

[20] Y. Mutoh and D. G. Sarvate, A 2×2 grid design, *Congr. Numer.*, 149 (2001) 193-199.

[21] P. A. Pevzner, *Computational Molecular Biology: an Algorithmic Approach*, MIT Press, Massachusetts, 2000.

[22] R. M. Phatarfod and A. Sudbury, The use of a square array scheme in blood testing, *Statistics in Medicine*, 13 (1994) 2337-1343.

[23] D. Raghavarro, *Constructions and Combinatorial Problems in Designs of Experiments*, Wiley, New York, 1971.

10
Non-unique Probe Selection

A probe is a short oligonucleotide of size 8-25, used for identifying viruses (or bacteria) in a biological sample through hybridization, where a biological sample is a biological object, such as blood, containing a subset of viruses. When each probe hybridizes to a unique virus, identification is straightforward. However, finding unique probes is often not so easy. In this chapter, how to select non-unique probles to identify viruses is introduced with group testing techniques.

10.1 Introduction

Although temperature and salt concentration are very helpful for a probe to hybridize uniquely to its intended virus, finding unique probes for given set of viruses is still a difficult task, especially for closely related virus subtypes. Schilep, Torney and Rahman [5] proposed a group testing method to use non-unique probes to find which viruses are in a biological sample. They consider a set of viruses hybridized to a probe as a pool and the incidence matrix of probes and viruses as a pooling design.

A pool yields a positive outcome if it hybridizes to a virus in the biological sample. The problem here differs from those in previous chapters in the fact that the candidate set of probes is more or less determined, not subject to our design as in previous chapters. What we can do is to construct a pooling design by selecting a subset of the candidate set.

If the incidence matrix is $\bar{d}$-separable, then the presence of up to d viruses in a given biological sample can be determined with those non-unique probes. Their methodology contains three steps:

Step 1. Find a large set of non-unique probes.

Step 2. From the set of probes obtained in Step 1, find a minimum subset of probes to identify up to d viruses.

Step 3. Decode the presence or absence of viruses in the given biological sample from testing outcomes.

In Step 2, they suggested a greedy algorithm which adds probe one by one until the incidence matrix with considered viruses form a $\bar{d}$-separable matrix. Actually, finding a minimun solution could be quite hard. Here, we discuss two problems:

> SEPARABILITY*-TEST. Given a binary matrix M and a positive integer d, determine whether M is $\bar{d}$-separable.
>
> MIN-$\bar{d}$-SS (Minimum $\bar{d}$-Separable Submatrix). Given a binary matrix M, find a minimum $\bar{d}$-separable submatrix with the same number of columns.

Consider the computational complexity of decoding. Thai, Deng, MacCallum and Wu [6] suggested to use d-disjunct matrix instead of $\bar{d}$-separable matrix and considered the following corresponding problems:

> DISJUCTNESS-TEST. Given a binary matrix M and a positive integer d, determine whether M is d-disjunct.
>
> MIN-d-DS (Minimum d-Disjunct Submatrix). Given a binary matrix M, find a minimum $\bar{d}$-disjunct submatrix with the same number of columns.

Clearly, the following two problems are closely related to the above four problems.

> SEPARABILITY-TEST: Given a binary matrix M and a positive integer d, determine whether M is d-separable.
>
> MIN-d-SS (Minimum d-Separable Submatrix). Given a binary matrix M, find a minimum d-separable submatrix with the same number of columns.

What is the power of non-unique probes? Is there a limitation for usage of non-unique probes? The following theorem says that there is a limitation, however, not very serious. That is, no matter how many non-unique probes are used, without using unique probes (singleton pools), we cannot identify all possible sets of viruses. However, if a sample contains at most $n-2$ viruses where n is the total number of considered viruses, then it is possible to identify them without using unique probes.

Theorem 10.1.1

(a) A $t \times n$ binary matrix is $\bar{n}$-separable if and only if every column is isolated.
(b) A $t \times n$ $(n-1)$-separable matrix contains at least $n-1$ isolated columns.
(c) There exists a $\overline{(n-2)}$-separable matrix without isolated columns.

Proof. (a) For every column i, consider samples $[n]$ and $[n] - \{i\}$. To distinguish them, we must use pool $\{i\}$. Therefore, column i is isolated. Conversely, if every column is isolated, then the matrix is clearly $\bar{n}$-separable.

(b) For any two distinct columns i and j, consider samples $[n] - \{i\}$ and $[n] - \{j\}$. To distinguish them, either pool $\{i\}$ or $\{j\}$ must be used. Therefore, either column i or column j is isolated. This implies that at least $n - 1$ columns are isolated.

(c) Consider all pools of two viruses. Row vectors corresponding to them form an $(n-2)$-disjunct matrix, hence $\overline{(n-2)}$-separable. □

In the remaining sections of this chapter, we present some results and our thoughts about computational complexity and approximation solutions of the above six problems.

10.2 Complexity of Pooling Designs

In this section, we study the complexity of SEPARABILITY-TEST, SEPARABILITY*-TEST and DISJUNCTNESS-TEST.

Theorem 10.2.1 SEPARABILITY*-TEST *is co-NP-complete.*

Proof. To show SEPARABILITY*-TEST is in co-NP, we can guess two samples from space $S(\bar{d}, n)$ and check whether M gives the same test outcome on the two samples.

To show the co-NP-completeness of SEPARABILITY*-TEST, we reduce the VERTEX-COVER to the complement of SEPARABILITY*-TEST by a method given in [1].

Consider a graph $G = (V, E)$ and a positive integer h $(1 \le h < |V| - 1)$. Let $V = \{1, 2, ..., m\}$ and $E = \{e_1, ..., e_k\}$. Every edge e_i is considered as a subset of two vertices. Without loss of generality, assume G has no isolated vertices. Set $d = h + 1$, we define an $(m + k) \times (m + 1)$ binary matrix M with rows corresponding to the following pools:

$$\begin{aligned} P_i &= \{i\} \text{ for } i = 1, 2, ..., m, \\ P_{m+i} &= e_i \cup \{m+1\} \text{ for } i = 1, 2, ..., k. \end{aligned}$$

First, assume G has a vertex cover C with size at most h. Define two samples

$$\begin{aligned} s_1 &= C, \text{ and} \\ s_2 &= C \cup \{m+1\}. \end{aligned}$$

Clearly, $s_1, s_2 \in S(\bar{d}, m+1)$ and $s_1 \neq s_2$. Let $P_i(s_j)$ denote the test-outcome of pool P_i on sample s_j. Then we have

$$\begin{aligned} P_i(s_1) = 0 = P_i(s_2) &\quad \text{for } i \in V \setminus C \\ P_i(s_1) = 1 = P_i(s_2) &\quad \text{for } i \in C \\ P_{m+i}(s_1) = 1 = P_{m+i}(s_2) &\quad \text{for } i = 1, 2, ..., k. \end{aligned}$$

Therefore, M is not $\bar{d}$-separable.

Conversely, assume that M is not $\bar{d}$-separable. Then there exist two samples $s_1, s_2 \in S(\bar{d}, m+1)$, $s_1 \neq s_2$ such that for all $i = 1, 2, ..., m+k$, $P_i(s_1) = P_i(s_2)$. Since $P_i(s_1) = P_i(s_2)$ for $i = 1, 2, ..., m$, we have $s_1 \cap V = s_2 \cap V$. Note that $m+1$ is the only item not in V. To have $s_1 \neq s_2$, $m+1$ must belong to exactly one of s_1 and s_2. Without loss of generality, assume $m+1 \notin s_1$ and $m+1 \in s_2$. Set $C = s_1 \cap V = s_2 \cap V$. Then $s_1 = C$ and $s_2 = C \cup \{m+1\}$. Therefore, $P_{m+i}(s_1) = P_{m+i}(s_2) = 1$ for $i = 1, 2, ..., k$. This implies that C is a vertex cover of G. □

Next, we study the SEPARABILITY-TEST and DISJUNCTNESS-TEST.

Theorem 10.2.2 SEPARABILITY-TEST *is co-NP-complete.*

Proof. We show SEPARABILITY*-TEST $\leq^p_m$ SEPARABILITY-TEST where $\leq^p_m$ stands for polynomial-time many-one reduction (see [3]). For a $t \times n$ binary matrix M in an instance of SEPARABILITY*-TEST, we put a zero column to form a $t \times (n+1)$ binary matrix M'. By Lemma 2.1.6, M is $\bar{d}$-separable if and only if M' is d-separable. □

Theorem 10.2.3 DISJUNCTNESS-TEST *is co-NP-complete.*

Proof. Note that the union of columns in a sample s is the testing outcome on the sample s. In the proof of Theorem 10.2.1, set $d = h$ instead of $d = h+1$. Then $P(s_1) = P(s_2)$ means that colum $m+1$ is contained in the union of d columns in s_1 and hence it is the evidence of M not being d-disjunct. Therefore, we can use the same reduction to show that VERTEX-COVER $\leq^p_m$ DISJUNCTNESS. □

Based on definition of $\bar{d}$-separability (d-separability, d-disjunctness), a naive method to check whether a binary matrix is $\bar{d}$-separable (d-separable, d-disjunct) takes $O(n^d)$ time. From Theorems 10.2.1, 10.2.2 and 10.2.3, we see that it is unlikely to have any smart way to reduce this running time. Therefore, non-unique probes can be successfully used only in case of small d.

10.3 Complexity of Minimum Pooling Designs

For fixed small d, what is the computational complexity of MIN-d-SS, MIN-$\bar{d}$-SS and MIN-$\bar{d}$-DS? In this section, we answer this question.

Fisrt, we note that MIN-1-SS is exactly the same problem of Minimum-Test-Sets in [3] and its NP-hardness is already proved.

Theorem 10.3.1 MIN-1-SS *is NP-hard.*

Next, we modify the reduction used in the proof of the NP-completeness of Minimu-Test-Sets in [3] to show NP-hardness of MIN-$\bar{d}$-SS and MIN-$\bar{d}$-DS.

Theorem 10.3.2 MIN-$\bar{1}$-SS *is NP-hard.*

Proof. Note that a binary matrix is $\bar{1}$-separable if and only if all pools corresponding to rows of M satisfy the following conditions:

(a) Every column appears in at least one pool.

(b) For every two columns x and y, there exists a pool containing exactly one of x and y.

To prove NP-hardness of MIN-$\bar{1}$-SS, we will reduce a well-known NP-complete problem, 3-dimensional matching [3], to the decision version of MIN-$\bar{1}$-SS.

> 3-DM (3-dimensional matching). Give three disjoint sets X, Y and Z with the same cardinality m and a collection $\mathcal{C}$ of 3-subsets of $X \cup Y \cup Z$, each of which contains exactly one element of X, one element of Y and one element of Z, determine whether $\mathcal{C}$ contains a subcollection $\mathcal{C}'$ such that each element of $X \cup Y \cup Z$ appears in exactly one 3-subset in $\mathcal{C}'$. (Such a subcollection $\mathcal{C}'$ is called a *3-dimensional matching.*)

> Decision Version of MIN-$\bar{1}$-SS. Given a $t \times n$ binary matrix M and a positive integer h ($1 \le h \le t$), determine whether M contains a $h \times n$ submatrix being $\bar{1}$-separable.

Consider an instance of 3-DM, X, Y, Z and $\mathcal{C}$. We construct an instance of MIN-$\bar{1}$-SS as follows:

Set $n = 3m+3$, $t = |\mathcal{C}|+3$ and $h = 3+m$. Let us label columns of M by elements of $X \cup Y \cup \cup Z \cup \{a, b, c\}$ where $\{a, b, c\}$ is disjoint from X, Y and Z. We define a binary matrix M by giving the collection $\mathcal{P}$ of all pools corresponding to row vectors of M,

$$\mathcal{P} = \mathcal{C} \cup \mathcal{Q},$$

where

$$\mathcal{Q} = \{X \cup \{a\}, Y \cup \{b\}, Z \cup \{c\}\}.$$

First, suppose $\mathcal{C}$ contains a 3-dimensional matching $\mathcal{C}'$. Then $\mathcal{C}' \cup \mathcal{Q}$ would satisfy conditions (a) and (b) and hence it gives a $h \times n$ $\bar{1}$-separable submatrix.

Conversely, suppose M has an $h \times n$ $\bar{1}$-separable submatrix M'. Consider collection $\mathcal{P}'$ of all pools corresponding to row vectors of this submatrix M'. $\mathcal{P}'$ must satisfy conditions (a) and (b). Condition (a) forces $\mathcal{P}'$ to contain $\mathcal{Q}$ since a, b and c appear only in $\mathcal{Q}$ and exactly once. Set $\mathcal{C}' = \mathcal{P}' \cap \mathcal{C}$. Then $\mathcal{C}'$ contains exactly m 3-subsets. For any element $x \in X$, since condition (b) holds for x and a, $\mathcal{P}'$ must contain a pool $A \in \mathcal{C}$ such that $x \in A$. Therefore, x appears in $\mathcal{C}'$. Similarly, every element in Y and Z appears in $\mathcal{C}'$. Therefore, each element appears exactly once and $\mathcal{C}'$ is a 3-dimensional matching. □

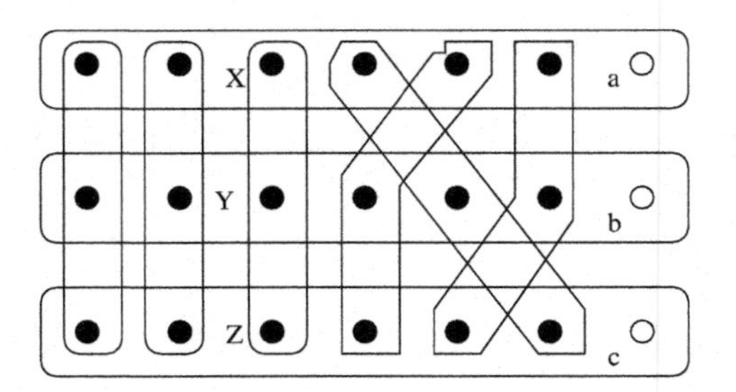

Figure 10.1: $\mathcal{P}'$

Theorem 10.3.3 MIN-1-DS *is NP-hard.*

Proof. Note that a binary matrix is 1-disjunt if and only if all pools corresponding to rows of M satisfy the following condition:

(*) For every two columns x and y, there exist two pools A amd B such that $x \in A$, $y \notin A$, $x \notin B$ and $y \in B$.

To prove NP-hardness of MIN-1-DS, we will reduce 3-DM to the decision version of MIN-1-DS.

> Decision Version of MIN-1-DS. Given a $t \times n$ binary matrix M and a positive integer h ($1 \leq h \leq t$), determine whether M contains an $h \times n$ 1-disjunct submatrix.

Consider an instance of 3-DM, X, Y, Z and $\mathcal{C}$. We construct an instance of MIN-1-DS as follows:

Set $n = 3m+3$, $t = |\mathcal{C}|+4$ and $h = 4+m$. Let us label columns of M by elements of $X \cup Y \cup \cup Z \cup \{a, b, c\}$ where $\{a, b, c\}$ is disjoint from X, Y and Z. We define a binary matrix M by giving the collection $\mathcal{P}$ of all pools corresponding to row vectors of M as follows:

$$\mathcal{P} = \mathcal{C} \cup \mathcal{Q},$$

where

$$\mathcal{Q} = \{X \cup \{a\}, Y \cup \{b\}, Z \cup \{c\}, \{a, b, c\}\}.$$

First, suppose $\mathcal{C}$ contains a 3-dimensional matching $\mathcal{C}'$. Then $\mathcal{C}' \cup \mathcal{Q}$ would satisfy condition (*) and hence it gives a $h \times n$ 1-disjunct submatrix.

Conversely, suppose M has an $h \times n$ 1-disjunct submatrix. Consider collection $\mathcal{P}'$ of all pools corresponding to row vectors of this submatrix. Since condition (*) holds for $\{a, b\}$, $\{b, c\}$ and $\{a, c\}$, $\mathcal{P}'$ must contain $\mathcal{Q}$. Therefore, $\mathcal{P}' \cap \mathcal{C}$ contains exactly m pools. Every element x of $X \cup Y \cup Z$ appears in $\mathcal{P}' \cap \mathcal{C}$ because conditon (*) holds for $\{x, a\}$, $\{x, b\}$ and $\{x, c\}$. □

From Theorems 10.3.1, 10.3.2 and 10.3.3, we may expect that MIN-d-SS, MIN-$\bar{d}$-SS and MIN-d-DS are NP-hard for every $d \geq 1$. However, we do not have a formal proof for $d > 1$.

10.4 Approximations of Minimum Pooling Designs

Klau, Rahmann, Schliep, Vingron and Reinert [3] proposed to find minimum $(\bar{1}; z)$-separable submatrix in a given binary matrix M, using integer linear programming. Let m_{ij} denote the entry in cell (i, j) of M. Then the problem is equivalent to the following:

$$\begin{array}{lll} \min & \sum_{i=1}^{t} x_i & \\ s.t. & \sum_{i=1}^{t} m_{ij} x_i \geq z & \text{for } 1 \leq j \leq n, \\ & \sum_{i=1}^{t} |m_{ij} - m_{ik}| x_i \geq z & \text{for } 1 \leq j < k \leq n, \\ & x_i \in \{0, 1\}, & \end{array}$$

where $x_i = 1$ if and only if the ith row is chosen to put in the submatrix. They studied a branch and bound approach to solve this integer linear programming.

Actually, this integer linear programming is a typical general covering problem. We can find from [2] two standard approximations. One is a greedy algorithm which produces an approximation solution within a factor of $1 + \ln(\frac{n^2}{4} + \beta)$ from optimal where β is an upper for row weights of input binary matrix M. Another one is a prime-dual algorithm which produces an approximation solution within a factor of α from optimal where α is an upper bound for both column weights and the Hamming distance of two columns in M. However, no smart approximation to this specific integer programming has been found.

References

[1] D.-Z. Du and K.-I Ko, Some completeness results on decision trees and group testing, *SIAM Algebraic and Discrete Methods*, 8 (1987) 762-777.

[2] D.-Z. Du and K.-I Ko, *Design and Analysis of Approximation Algorithms*, manuscript, 2006.

[3] M.R. Garey and D.S. Johnson, *Computers and Intractability*, (W.H. Freeman, San Francisco, 1979).

[4] G. Klau, S. Rahmann, A. Schliep, M. Vingron, and K. Reinert, Optimal robust non-unique probe selection using integer linear programming, *Bioinformatics*, 20 (2004) I186-I193.

[5] A. Schliep, D. C. Torney and S. Rahmann, Group testing with DNA chips: generating designs and decoding experiments, *Proceedings of the 2nd IEEE Computer Society Bioinformatics Conference*, 2003.

[6] M. Thai, P. Deng, D. MacCallum and W. Wu, Dealing with non-unique probes with d-disjunct matrices, manuscript, 2006.

Index